Gas Chromatography-Mass Spectrometry
How Do I Get the Best Results?

Gas Chromatography-Mass Spectrometry
How Do I Get the Best Results?

By

Diane C. Turner
Anthias Consulting Ltd, UK
Email: dct@anthias.co.uk

Mathias Schäfer
University of Cologne, Germany
Email: aco78@uni-koeln.de

Steven Lancaster
Domino Printing Sciences, UK
Email: Steven.Lancaster@domino-uk.com

Imran Janmohamed
Anthias Consulting Ltd, UK
Email: i44ran@hotmail.com

Anthony Gachanja
Jomo Kenyatta University of Agriculture and Technology, Kenya
Email: agachanjah@yahoo.com

and

Jason Creasey
GlaxoSmithKline, UK
Email: jason.m.creasey@gsk.com

ROYAL SOCIETY
OF **CHEMISTRY**

Print ISBN: 978-1-78262-928-3
EPUB ISBN: 978-1-83916-004-2

A catalogue record for this book is available from the British Library

The Royal Society of Chemistry is a charity, registered in England and Wales, Number
207890, and a company incorporated in England by Royal Charter (Registered No.
RC000524), registered office: Burlington House, Piccadilly, London W1J 0BA, UK,
Telephone: +44 (0) 20 7437 8656.

Visit our website at books.rsc.org

Preface

Chemical analysis is an important step in identifying problems over a wide spectrum of issues including environmental, medical, forensic, agricultural, transportation, industrial, planetary and space sciences research and development. Good analysis is vital for executive decision-making in all of the above fields. Without the appropriate chemical analysis, data analysis and interpretation, organisations and governments are denied critical information which results in decisions being made with a high probability of them being flawed. The analytical laboratory or analyst therefore needs to understand the nature of the problem at the earliest possible stage.

Analysis is expensive, but no analysis or poor analysis is even more costly. There is no purpose in spending resources to generate data which nobody will use or is inaccurate and/or imprecise. However, analysis that is targeted to address a well thought through problem or question and that has been well planned results in the consumer of this data gaining significant value and knowledge from it. The analytical laboratory and well-trained analyst therefore fills an important role in answering and solving societal problems all over the world. In order to ensure that this value is achieved, the educated and trained analyst will communicate with the users of his or her data to ensure that the correct questions are being asked and that the problem is well defined. Once the data has been generated, the analyst will then work with the user to ensure that maximum knowledge can be derived

Gas Chromatography-Mass Spectrometry: How Do I Get the Best Results?
By Diane Turner, Mathias Schäfer, Steven Lancaster, Imran Janmohamed, Anthony Gachanja and Jason Creasey
© Diane C. Turner, Mathias Schäfer, Steven Lancaster, Imran Janmohamed, Anthony Gachanja and Jason Creasey 2020
Published by the Royal Society of Chemistry, www.rsc.org

from the data. In this way, a good analytical facility with trained and motivated analytical scientists is critically important and can generate significant value.

Whenever analysis is required, the reasons for performing the analysis should be properly understood in order to acquire the correct data. Frequently, the first questions that must be answered are: "what is it?" (qualitative) and "how much is present?" (quantitative). Although the reader is encouraged to ask the question "why do you require this analysis?" This often leads to a better understanding of the problem and a much more satisfactory outcome. The analyst has at their disposal a range of tools (techniques) to answer the questions, varying in complexity and cost. The type of analytical method employed will depend on the:

- Nature of the analyte(s) of interest.
- Concentration ranges of the analyte(s).
- Matrix and potential interferences.
- Accuracy and precision required.
- Availability of equipment and reagents.

Introduction

Importance of Gas Chromatography-Mass Spectrometry (GC-MS) Analysis

GC-MS provides a versatile analytical technique for the analysis of chemical substances involved in virtually all sectors in economic and social development: trade, mineral exploitation, medicine, food, water and environment, to highlight just a few. It has the capacity to play a pivotal role in providing both qualitative and quantitative scientific data that will aid and catalyse development so that sustainable policies can be formulated and their implementation monitored. Provision of quality water is a major challenge in the developing world and so is the challenge of providing quality analytical data which will be directly addressed through an increase in the capacity in analytical chemistry. This will, as a result, create and sustain chemical monitoring and the management of activities. This is achievable by the provision of hardware (laboratories and equipment) but also in parallel, the training of scientifically qualified and practically trained human resources. Many countries are unfortunately poor in both aspects and GC-MS instruments are still scarce in these laboratories!

Gas Chromatography-Mass Spectrometry: How Do I Get the Best Results?
By Diane Turner, Mathias Schäfer, Steven Lancaster, Imran Janmohamed, Anthony Gachanja and Jason Creasey
© Diane C. Turner, Mathias Schäfer, Steven Lancaster, Imran Janmohamed, Anthony Gachanja and Jason Creasey 2020
Published by the Royal Society of Chemistry, www.rsc.org

For example, in Africa activities are diverse, including transportation, mining, manufacturing, food and agriculture, drugs and pharmaceuticals, tourism and wildlife. The identification of materials for the processing industries impacts on the custom duty revenues. Similarly, any suspected drugs of abuse, for example cocaine need to be identified. This leads to forensic laboratory capacity for which MS data will lead to success in the prosecution of offenders. The same facility will find applications in the detection of both human and wildlife chemical poisoning, for the identification of chemicals used, which ultimately leads to policy adjustments/formulation to reduce the possibility of chemical misuse. The agricultural sector uses chemicals for pest control, including animal pest and disease control. GC-MS should be the technique of choice for the identification of organic chemicals which are volatile at the instrument operating temperatures.

For the safety of food, destined for both the local and export market, regular surveillance of chemical residues is very important. The pharmaceutical industrial sector is growing rapidly in Africa. Although the main equipment used in their laboratories are UV-visible and infrared spectrophotometers alongside liquid chromatographs, some pharmacopeia protocols require GC analysis. The industry would benefit from access to GC-MS, particularly with headspace sampling for conducting analysis, for example, of residual solvents in raw materials. GC-MS is an important tool for the testing of materials, maintenance of standards and specifications of products.

The areas of pollution and environmental contamination of air, water and soil would benefit significantly from GC-MS analysis. Recently the occurrence of cancer related mortalities in the region has greatly increased. The relevant laboratories in the areas of the world with the mandate for environmental pollution analysis do not have GC-MS and LC-MS capabilities. The identification of organic pollutants in the environment within Africa currently remains a significant challenge.

Crude oil has recently been discovered within the African region and commercial drilling and exploitation are ongoing or expected to commence soon. This will result in increased potential environmental pollution and the development of related industries creating further potential issues. GCMS is a vital tool to the petroleum and petrochemicals industries.

Provision of instrumentation without the consideration of maintenance, spare parts, consumables, power stabilisers and back-up solutions, only results in instruments in institutional laboratories gathering dust, constantly devaluing and being of no benefit to anybody. Instruments will continue to be unavailable to benefit the region if there is no well thought-out plan for procurement and importation, installation and qualification, servicing and maintenance and finally decommissioning. Locally available and institutional staff with a reasonable skill level in GC-MS maintenance must be available for the instrument to have the expected impact on society. GC-MS training workshops and other training for scientists and technical staff will go a long way to address this issue.

Applicability of GC-MS

Gas chromatography is an analytical separation technique, used to separate and detect chemical components in a sample mixture, which are usually organic compounds and gases, otherwise known as analytes, with molecular weights below 1250 u. A column is used to perform the separation, with the mobile phase being a gas, while the stationary phase may be a solid or a liquid coated on the inside of the column or held (immobilised) on a solid support packed into the column. The sample components need to be in the gas phase during the chromatographic separation process, as components will preferentially elute through the column at a rate determined by their affinity for the stationary phase. The requirements for analytes to be analysable using GC are that they:

- Must be volatile enough to be vaporised and move through the GC at operational temperatures below 480 °C, although most common GCs do not operate above 350 °C. Some analytes can be modified to form volatile derivatives that are GC amenable.
- Be thermally stable and not decompose at the temperature required to vaporise the sample.
- Should not undergo reactions with the stationary or mobile phases or be broken down by any flow-path components during the analysis.
- Should give a signal with the detection system and be visible above the limit of detection of the method at the lowest concentration required.
- Interact reversibly with the stationary phase and not be irreversibly adsorbed in the system.

The role of chromatography is to separate the components of a complex mixture while the detection system gives a signal as each of the components elute from the column. In cases in which the above mentioned conditions cannot be satisfied, then gas chromatographic analysis will not be the best analytical technique. Other techniques include liquid chromatography, spectrophotometric, isotopic, nuclear and electroanalytical analyses.

Problem analytes include those that are thermally labile, have a high molecular weight, are present at trace levels or are active in the system, being irreversibly adsorbed or broken down thermally or catalytically. Some of these analytes may still be analysed using GC, however more care is needed when choosing, optimising and maintaining the analytical system. For example, thermally labile analytes could be introduced using a cold injection technique. Explosives are analysed by optimising the method with temperatures as low as possible for each parameter and ensuring the system is as clean as possible to prevent activity.

Available detection systems for GC include selective and non-selective detectors, of which there are many! Non-selective detectors respond to specific bulk properties of the molecules, for example thermal conductivity. Selective detectors respond to specific properties unique to a class of molecules, such as specific elements, bonds, functional groups, electron capture capability, the combustion of organic molecules to give specific radiation or ions with a mass to charge ratio. All detectors have varying sensitivities and dynamic ranges. Universal detectors are good for observing most organic compounds in a sample, whereas selective detectors can reduce matrix interferences and improve detection limits.

Mass spectrometry is an analytical technique that can be used to identify unknown analytes, quantify known analytes and determine the structural and chemical properties of molecules, usually organic. The mass selective detector (mass spectrometer) gives the mass to charge ratio of ions produced by the molecules eluting from the GC column. This means that the two systems need to be coupled or interfaced so that each can operate under optimal conditions of pressure and temperature. The mass spectrometer operates under reduced pressure (vacuum) while GC separation takes place under pressures higher than atmospheric pressure. The mass spectrometer is comparatively more expensive to acquire than the flame ionisation (FID), thermal conductivity (TCD) or electron capture (ECD) detectors. However, the mass spectrometer has the advantage of unequivocal identification of the molecules as they elute from the column. In some cases, the molecular ion is not detected because it fragments as soon as it is formed. The fragmentation pattern of a compound is unique, like a fingerprint. It may be used for identification of the analyte through mass spectral interpretation and a library of mass spectra can be built. The availability of library spectra is a useful tool in qualitative analysis for the fast identification of analytes through library searching. Only a fraction of the molecules that enter the ion source in the mass spectrometer are converted into ions. Compared to selective detectors, for example ECD, only the most modern MS instruments can now approach the lower detection limits for some analytes. It is also important to note that the MS does not distinguish between many isomers and therefore the retention time of the analyte on the GC column, if the isomers have been separated chromatographically, is important for identification.

This Book

This book is not a book on mass spectrometry, there are plenty of those out there which you may like to reference. Nor is it a book on GC, sample prep or analytical statistics – there are plenty of these out there too. This is a book written around the concept of the "How do I...?" question, taking the reader from the start of their analysis to the final conclusion. The start of any method is the sample, thinking about how it is collected, transported to the laboratory and stored, ready for analysis. Then, the analysis itself including how to prepare the sample, how to introduce the sample onto the GC column, how to separate the components and how to detect those components. The final step is data analysis, how to find out what the components are, how much of each component is present and report the findings with confidence. Underlying all of this is the GC-MS instrument itself which needs maintenance and troubleshooting to keep it fully operational and to keep giving the correct answers. So how do I select the best techniques for my application? How do I develop and optimise my method so that it is robust and gives the correct answers? How do I look after my instrument?

This book has been written by authors with a significant amount of experience of performing and teaching GC-MS all over the world, including in Africa, and who understand the challenges that analysts face when it comes to solving problems using analytical science. It should be a useful guide for anyone who is called upon to solve problems *via* the application of GC-MS in any country. It will help the reader to develop their GC-MS knowledge in a practical way and not just learn how the techniques work but how to get their instrument working with high quality, robust methods and keep it working.

Contents

Gas Chromatography-Mass Spectrometry: How Do I Get the Best Results?
By Diane Turner, Mathias Schäfer, Steven Lancaster, Imran Janmohamed, Anthony Gachanja and Jason Creasey
© Diane C. Turner, Mathias Schäfer, Steven Lancaster, Imran Janmohamed, Anthony Gachanja and Jason Creasey 2020
Published by the Royal Society of Chemistry, www.rsc.org

1 Sample Collection and Preparation: How Do I Get My Sample Ready for GC-MS Analysis?

In order to conduct a successful GC-MS analysis there are some fundamental requirements. Gas chromatography relies on the analytes being in the gas phase; therefore, one important element is to ensure the optimal conditions are in place for this to occur. The second fundamental part is ensuring that a sufficient amount of the analyte reaches the detector. Although a mass spectrometer is a very sensitive detector, like every detector, it has limits. Therefore, any successful sample preparation must consider these limitations.

There are many different sample preparation techniques which can be used or automated for GC-MS. The best sampling or sample preparation technique to use is determined by:

- Sample phase: gas/liquid/solid or something in-between?
- Where is the sample? Can a portion be moved into the lab or must it be sampled *in situ* (can the instrument be taken to it)?
- Analytes: volatile/semi-volatile/involatile?
- Is it possible, if necessary, to automate the sampling/preparation technique?

Gas Chromatography-Mass Spectrometry: How Do I Get the Best Results?
By Diane Turner, Mathias Schäfer, Steven Lancaster, Imran Janmohamed, Anthony Gachanja and Jason Creasey
© Diane C. Turner, Mathias Schäfer, Steven Lancaster, Imran Janmohamed, Anthony Gachanja and Jason Creasey 2020
Published by the Royal Society of Chemistry, www.rsc.org

A sample will be in one of three states, solid, liquid or gas. The technique used for its successful introduction into the GC-MS will vary based on the sample state. This chapter is therefore sub-divided into these three states.

1.1   How Do I Collect and Sample a Gas for GC or GC-MS Analysis?

Gas-phase samples are already in the state in which GC separations occur, therefore, there is no need for further transformation. Some gas-phase samples must be sampled *in situ*, for example air, breath from a patient or air from a processing plant. For others, a large sample can be taken and then a portion of this analysed, for example a cylinder of industrial gas, or a canister filled with the gas-phase sample.

The analytes in gas-phase samples are usually already gaseous, therefore the sampling of gases for analysis should be quite straight forward, as long as the sample is kept under leak-free conditions. However, even samples of this type have some challenges, these include:

- Enrichment of the sample to ensure the concentration is high enough to allow successful detection.
- Accurate sampling to ensure the sampled fraction is representative of the bulk.
- Transfer to the GC-MS system which does not change the sample through reaction or absorption onto surfaces and ensures the sample volume entering the GC is optimal for the gas chromatographic process, for example, delivered in a narrow band onto the head of the column.
- Storage of the sample before analysis to maintain its integrity both qualitatively and quantitatively.
- The preparation of suitable standards for quantitative analysis.

Sampling an accurate representation of the bulk is a key step. The gaseous environment can be sampled in a variety of ways and these are described in a number of standards, for example ASTM (D3588, D5466), EPA (TO-14A & TO-15, EPA Method 18) and ISO (3171).

As these standards indicate, there are a wide variety of methods used to sample a gas and it usually takes place in three different ways: spot, continuous or representative.

1.1.1   What Is Spot Analysis?

Spot samples are taken at one time and at one point, usually *via* a pitot tube. The pitot tube is a pressure measurement instrument and its primary function is to measure the fluid flow velocity of liquid, air and gas flows. It is inserted into the process (*e.g.* stack monitoring) or pipeline (*e.g.* gas outlet), or *via* a valve. Analysis occurs immediately without the need for storage.

1.1.2   How Do I Sub-sample with a Canister or Sampling Bag?

There are various methods that can be employed to fill and use canisters and sampling bags, a full description of which is beyond the scope of this chapter. They all share the aim of creating a representative sample from the bulk being sampled. They all involve adding a flow controller to a sample over extended periods of time rather than immediate sampling.

For example, GPA Standard 2166 describes eight different sampling methods which are listed below:

- Evacuated container method: gas is introduced into an evacuated sample container with a pressure of less than 1 mm Hg.
- Reduced pressure method: similar to the evacuated contained method, but for higher inlet pressures.
- Helium pop method: beginning with an evacuated sample container, this is filled with helium (to around 5 psi), and then filled with the gas sample.
- Floating piston cylinder method: this method has a pre-charge chamber and sample chamber created by a piston. The pre-charge chamber is filled with an inert gas (slightly above line pressure). The outlet valve is opened and the sample displaces the piston and fills the cylinder.
- Water displacement method: the sample cylinder is filled with clean water and a vessel to measure the displaced water is

attached. The gas sample is slowly introduced and the outlet valve slowly opened. The gas is sampled until all the water is displaced (detected using the sound or by observation).

- Glycol displacement method: the same as the water displacement method but using glycol rather than water.
- Purging – fill and empty method: the sample is used to purge the container, it is then emptied by releasing the output valve. This process is repeated several times to obtain a representative sample.
- Purging – controlled rate method: the rate of entry is controlled by flow controllers on the inlet and outlet.

1.1.2.1   *How Do I Select and Use a Canister?*

Deactivated SUMMA™ or SilcoSteel™ canisters have been internally treated to ensure the collected analytes do not react with the stainless steel surface. SUMMA™ canisters use a passivation process to apply a nickel-chromium oxide layer, whereas SilcoSteel™ canisters have an internal silica layer. However, the term SUMMA™ canister is sometimes used to describe both types of coated canister (ASTM D5466 – 15).

When selecting a canister, the volume and nature of the sample, and the location that the sample is being taken needs to be taken into consideration.

1.1.2.2   *How Do I Select and Use a Gas Sampling Bag?*

Before sampling, unused bags should be stored in a clean environment and sealed in an outer bag to prevent adsorption of contaminants. Bags should be pre-cleaned before use by flushing with high-purity nitrogen. For validation, compounds must be stable in the bag or canister over the period in which the validation is conducted. Overall, the leak rate from the bag must be low.

During sampling it should be ensured that any tubing used for the bag connections is clean. A known and predictable flow rate should be used. Bags should not be overfilled, no more than 80% of the stated maximum bag volume should be filled.

Bags are intended for a single use, owing to potential sample adsorption onto the bag film. Therefore, after sampling and analysis, best practice is to not re-use bags. Hold times in bags before analysis are typically recommended to be 48 h or less owing to concerns around adsorption onto bag surfaces, unless the validation study demonstrates a longer stability. Bags containing samples should be protected from direct sunlight and stored above 0 °C to prevent condensation. Bags should be transported in rigid containers to prevent bag puncture and not shipped by air unless samples will be kept in a pressurised area.

1.1.3 What Is Active Sampling?

Gas-phase samples can be actively sampled *in situ* by drawing the sample through a conditioned sorbent and packed into a thermal desorption (TD) tube using a constant pressure or constant flow pump. The tube is then sealed, to prevent the loss of analytes and the ingress of contaminants, then returned to the lab for analysis. Tubes can be stable for several weeks. For example, parts-per-trillion (ppt) levels of polyaromatic hydrocarbons (PAHs) can be detected using air analysis, by drawing 100 L of air through a packed TD tube. See Section 1.3.1 for background information on thermal desorption.

The solid sorbent can also be held within a device such as an ORBO™ tube or filter. An impinger enables collection with a liquid sorbent.

1.1.3.1 What Is Thermal Desorption?

Thermal desorption (TD) is a physical separation process, in which heat is applied to a sample to transfer analytes that are adsorbed or absorbed within the sample tube, into the gas-phase so that they can be analysed using gas chromatography. TD only uses temperatures up to 350 °C and therefore no chemical bonds are broken in the process, only interactions. TD can be used to analyse a range of species, from those as volatile as acetylene (with two carbon atoms) up to molecules with forty carbons, such as PAHs and phthalates.

Small solid or viscous liquid samples can be directly thermally desorbed by placing them in a conditioned TD tube (see Section 1.3.1). As mentioned previously, gas-phase analytes can be concentrated by drawing the gas-phase sample through a conditioned TD tube packed with a sorbent.

1.1.3.2 How Do I Select My TD Tube and Sorbent to Trap My Gas-phase Analytes?

The TD tube itself can be made from: glass, which is beneficial for observation of the position of solid samples placed directly into the tube (see Section 1.3.1); stainless steel, which makes it very robust, especially for those tubes sampled away from the lab; or coated (silco) steel, which makes the tube very inert and is much better for active analytes such as those molecules containing sulphur.

The tubes vary in size depending on the manufacturer, but the industry standard TD methods, such as ISO, CEN, ASTM and EPA, use a 3.5 × ¼ in. outside diameter (o.d.) tubes. Tubes should have a unique identifier, which enables the sample to be matched to the tube and the sampling direction must be known so that the tube is desorbed in the reverse direction to ensure that all the sampled analytes are recovered.

Similar to solid phase micro-extraction (SPME), the sorbents placed into the TD tube and the cold trap can either interact with the analytes through absorption or adsorption. The sorbent(s) selected is dependent on the target analytes. It must trap the target analytes at the ambient temperature of the sampling location and easily release them again when rapidly heated. This temperature must not be higher than the maximum temperature of the sorbent, with no irreversible ad/absorption or catalytic breakdown.

Common sorbents are polymers such as Tenax ®, Porapak, Hayesep or Chromosorb, a styrene divinylbenzene (DVB) polymer; carbon molecular sieves such as Sulficarb, Carbosieve or Carboxen; zeolite molecular sieves; or graphitised carbon black such as Carbopack, Carbotrap or Carbograph. Tenax® and graphitised carbon blacks are hydrophobic and are therefore beneficial for 'wet' samples. Carbon molecular sieves are mostly hydrophilic with Carboxen being the most hydrophobic. Zeolite molecular sieves are very hydrophilic and can collect water up to the mg level, in a typically sized TD tube.

Different types of sorbents are good for different volatilities and polarities of analytes and have different retention volumes and maximum temperatures (ranging from 190–400 °C). Even when the maximum temperatures are not exceeded, some sorbents can produce artefacts, that is the release of molecules, that are focused, separated and detected using GC. Carbon molecular sieves have minimal artefact levels, whereas Tenax® has a low artefact level when new, but the artefact level increases as the sorbent ages.

Porous polymers and carbonised molecular sieves are more inert than graphitised carbon blacks such as Carbograph 1TD. Generally, the more volatile the analyte(s), the stronger the sorbent must be. For analytes with a boiling point (bp) greater than 100 °C a weak sorbent such as Tenax® TA is used. Those analytes with a boiling point between 30–100 °C require the use of a medium strength sorbent such as Carbograph 1TD. Very volatile analytes with a boiling point between 30–50 °C require the of use a strong sorbent such as Sulficarb or Carboxen 1000.

The mesh size of the packing material affects the packing density and the back pressure that it creates. A mesh size of 20/40 has larger particles than 60/80 and therefore can be sampled using higher flow rates.

The sorbent life of the tube is dependent on the type of sorbent(s) used, the maximum and routine desorption temperatures that it has been exposed to and the number of desorption cycles, which includes conditioning of the tubes. Tenax® and carbon-based sorbents are usually good for 100–200 cycles, whereas porous polymers are less stable, usually with a lifetime of 100 cycles.

A trap is used to collect and focus analytes between the TD tube and the GC analytical column. Without this, the long TD tube desorption times would result in long transfer times to the column, producing broad sample bands and poor chromatographic resolution and peak shapes. The analytes can be trapped through cryofocusing in the inlet liner or on a GC pre-column using cryogens or through cold trapping with a Peltier cooled trap (cold trap). The cold trap enables the use of a small amount of sorbent in a narrow tube to selectively concentrate the analytes, but unlike the TD tube, the cold trap is cooled below the ambient temperature, reducing the likelihood of break-through even though less sorbent is used. With cryofocusing, everything released from the TD tube is trapped, but with the cold trap the sorbent can be selected so as not to trap unwanted gases such as water, solvents, and so forth. As with the TD tube sorbent, it must trap the target analytes at the (lower) temperature chosen and then easily release them with no catalytic breakdown when the trap is rapidly heated. As the cold trap is usually backflushed, multiple sorbents can be chosen, sometimes up to three, to match the target analytes. Upon rapidly heating the narrow cold trap, the analytes are typically transferred to the GC column in split mode to increase the flow through the cold trap when desorbing, resulting in a faster transfer to the GC column and therefore producing a narrow sample band. Even with a low split ratio, the sample flow to waste is higher than that onto the column resulting in a fraction of the sample being separated on the GC. Recently developed instrumentation can automatically enable the split flow effluent from the cold trap desorption to pass back through the original sample TD tube or through a new, conditioned TD tube to re-collect the sample. This means that TD samples are no longer one-shot, where if something went wrong with the analysis of the sample it could not be re-analysed. In addition, the recovery is quantitative, therefore if the sample does have to be re-analysed the original concentration can be determined based on the split ratio. Sample re-collection can also be achieved through the trapping of the split effluent from the TD tube desorption onto a new, conditioned TD tube if a split method was used in that step.

1.1.3.3   How Do I Desorb My TD Tube for GC-MS Analysis?

After placing the sample in the tube or, concentrating the gas-phase sample in a packed tube, the tube is placed in the TD instrument and checked for leaks. Upon passing, it is pre-purged with a carrier gas, usually helium, to remove atmospheric oxygen and to prevent oxidation of the sample upon heating. The TD tube is then heated to a temperature dependent on the maximum temperature of the sorbent, the volatility of the analytes and the nature of the sample. It is then held at that temperature, usually for 5–30 minutes, to fully desorb the sample or sorbent. The TD tube is continuously purged while being heated and the emerging analytes are back-flushed off the TD tube and selectively concentrated on a narrow cold (focusing) trap, containing a small amount of sorbent. During this process, unwanted gases, water and matrix are removed. If the sample is very concentrated, the analytes can be transferred to the cold trap in the split mode, rather than the usual splitless mode. After the TD tube is fully desorbed, the cold trap is rapidly heated and the concentrated analytes are backflushed through a heated transfer line, usually as a split injection, onto the GC analytical column in a narrow sample band for separation and detection.

1.1.3.4   What Other Parameters Do I Need to Consider for
Thermal Desorption?

Before use, TD tubes must be conditioned and sealed before being taken to the sampling location. Those packed with a sorbent must be conditioned at the optimal temperature for the required time for the specific sorbent(s) used.

The break-through volume for the samples and analytes to be sampled must be determined before sample collection, so that a known volume of the sampled gas can be taken which is below the break-through volume. The break-through volume can be simply checked by attaching two sampling tubes in series; if the first tube becomes saturated any analytes breaking-through will be trapped by the second tube, this can then be analysed to determine if the break-through volume has been exceeded.

During sampling, the optimal flow rate through the tube affects the interaction of the analyte with the sorbent and therefore the amount that can be trapped. The optimal sampling rate through a standard 5 mm inside diameter (i.d.) packed TD tube is 50 mL min^{-1}, with a working range of 10–200 mL min^{-1} that can increase to 500 mL min^{-1} for a maximum of 10 to 15 min. Sample volumes range from 500 mL to 100 L.

Improved adsorption efficiencies are observed at lower temperatures, however the ambient temperature during sampling must be considered when selecting the sorbent(s). If there is a wide range of physical and chemical properties of the analytes, a tube can be packed with multiple sorbent beds, usually up to a maximum of three, so that there is a sufficient mass of each sorbent for the capacity required. The weakest sorbent is packed closest to the sample, with the strongest sorbent packed furthest from the sampling end of the tube to enable the fast and efficient release of the analytes in the reverse direction. The analytes are then back-flushed to the next stage of the TD process.

If replicate sample aliquots are required, a manifold sampling system can be used in which there are multiple tubes, either containing the same sorbent(s) for repeatability checks or containing different sorbents to sample different analytes. These are set-up in parallel and are all connected to the same pump, but with flow control valves on each, to either balance the flows or to set individual flows for each TD tube.

Quantitation can be performed, as with other techniques (see Chapter 8), by analysing the standards using the same method as the samples. A maximum of 1–2 μL of standard solution, preferably in a solvent that is not trapped by the sorbent and has a low vapour expansion coefficient, or its headspace is directly spiked onto the TD tube using a spiking rig. The rig enables the standard to be injected into a carrier gas so that the analytes are blown through the TD tube sorbent to be trapped in the same manner as the sample analytes and then any solvent is purged before analysis. The automatic addition of an internal standard (IS) into the TD tube and/or the cold trap is possible and is instrument dependent.

1.1.4   What Is Passive Sampling?

Passive sampling is achieved when contaminated air enters the device and diffuses through the sorbent without the use of any pumps. The sorbent can then be analysed by thermally desorbing the analytes into a GC-MS system, for example *via* a TD instrument or with liquid desorption and injection.

Passive sampling is frequently used for occupational health to monitor indoor and ambient air. A packed TD tube with an axial diffusive sampler is used for long-term exposure, from 8 hours to 4 weeks. Radial diffusive samplers have a much larger surface area and therefore are good for short-term monitoring of 0.5 to 6 hours. After sampling, the cartridge is removed from the sampler and placed in an empty TD tube for analysis. Monitoring badges are similar to radial diffusive samplers. Standard methods present compound-specific diffusion uptake rate data for different devices, which is used to quantify the concentration of analytes in the air.

1.1.5   What Is Online Sampling?

Thermal desorption instruments can also be configured for online sampling, in which the gas-phase analytes are trapped, concentrated and analysed. The packed TD tubes themselves are not needed, as the sample is pumped directly through the cold (focusing) trap. Two traps in parallel enable continuous monitoring and they can operate, unattended, for days, weeks or even years! These are commercially available from a number of suppliers including Markes International or Perkin-Elmer.

1.2   How Do I Sample and Prepare a Liquid for GC or GC-MS Analysis?

Some liquid-phase samples may need to be sampled *in situ*, for example process and waste streams. For others, a sub-sample may be taken and then a portion of this analysed in the lab, for example water, chemicals, urine or fuel. The analytes in liquid-phase samples can range from volatile to involatile and therefore care must be taken not to lose these when sampling, transporting and storing.

As with a gaseous sample the same key points need to be considered:

- Storage of the sample before analysis to maintain its integrity both qualitatively and quantitatively.

- Dilution or enrichment of the sample analytes to ensure concentration is high enough to allow successful detection, the concentration should be in the working range of the GC-MS method. Lower or higher analyte concentrations can be obtained through several stages of the analysis process:
  - During sample preparation through dilution with a solvent or by using a sample enrichment technique or solvent blowdown concentration step under nitrogen.
  - By injecting a very small (*e.g.* 0.1 μL) or large volume (up to 1000 μL) of a liquid sample, see Chapter 2.
  - During sample introduction to the column by performing split or splitless injection, see Chapter 2.
- Transfer the sample into the GC system in a way that does not change the sample through reaction or adsorption onto surfaces and ensures the sample entering the GC is optimal for the gas chromatography process, that is, it is delivered as a narrow band onto the head of the column, see Chapter 2.

1.2.1 How Do I Store My Liquid Samples?

Successful storage of a liquid sample is very application dependent, however, broadly speaking the integrity of samples can be maintained by slowing or eliminating sources of chemical or biological degradation. This may include:

- Protection from light, to prevent photo-degradation.
- Protection from high temperatures to minimise evaporation of volatile analytes, thermal degradation and growth of microbes.
- Protection from the ingress of reactive components such as oxygen or moisture.

The extent of protection required is established on a case-by-case basis. However, for general protection a well-sealed container is used, purged if necessary with dry nitrogen to remove reactive oxygen and stored in a fridge, cupboard or other monitored environment.

1.2.2 Which Solvents Can I Use in Sample Preparation for Injection into a GC?

Whether diluting a liquid sample for direct injection, or using a solvent in the sample preparation process a solvent should be selected that is appropriate for both the sample and the analysis method:

- Choose a solvent which dissolves the analytes of interest but does not react with them. This probably means a solvent which is similar in nature to the sample. It should also be miscible with the sample solvent, otherwise this will result in a liquid–liquid extraction (LLE).
- Choose a solvent that is GC-amenable that:
  - Enables the GC to deliver sharp, well resolved peaks in which the solvent is either eluted quickly at the start of the chromatogram or eluted later after the peaks of interest are separated and eluted. The former is more common and enables easier method development than the latter arrangement, although for some methods the solvent elutes between the peaks but this should be avoided if possible.
  - Is suitable for the sample introduction technique being used. This includes having a boiling point (b.p.) that is relatively lower than the most volatile analytes to be determined. The lowest initial oven temperature that is attainable without cryogenic cooling in the environment in which the GC is sited should also be considered if solvent re-condensing on the column is required. For example, if dichloromethane (b.p. 39 °C) is used for an on-column, splitless or large volume injection, a low oven initial temperature at a maximum of 35 °C, preferably 30 °C, must be easily attainable. Another factor is the solvent polarity relative to the inlet liner packing material and column stationary phase, see Chapter 2 for more information on these points.
  - It should not react with any GC components, for example the injection syringe, inlet septum or liner, column stationary phase or detector. This can lead to leaks where it reacts with seals such as the septum, contamination producing additional peaks in the chromatogram, high bleed which produces high baselines, and can ultimately result in higher consumable costs, poor chromatography and sensitivity.

Therefore, typically the solvent is volatile with a lower boiling point than the most volatile analyte. This allows the widest range of options when setting parameters for the GC and also allows the start of the MS data collection to be delayed until after this solvent peak has cleared the vacuum system, this is called the solvent delay. The MS filaments within the ion source are turned off until after the solvent delay, to prevent them from being overloaded by this large quantity of solvent which will cause them to prematurely fail. The pressure in the mass spectrometer is also higher at this point which can affect the performance of the mass spectrometer if it is acquiring a spectrum, thereby reducing sensitivity. Information on how to optimise the solvent delay time is discussed in Chapter 5.

1.2.3 Which Sample Preparation Techniques Can Be Used for Liquid Samples with GC-MS Analysis?

Very few liquid (or solid) samples can be directly introduced into a GC without some form of preparation. The reasons for sample preparation depend on the nature of the sample itself and can include one or all of the following examples:

- Removal of interfering matrix components which could prevent the detection of the analytes.
- Protection of the analytical system from contaminants, which would cause frequent maintenance and high consumable costs.
- Selective isolation and concentration of the analytes with 100–5000x enrichment.
- Changing the phase or the sample solvent: for example, water samples are very difficult to directly inject.

The transition from the liquid to the gas phase places some limitations on the volume of sample which can be introduced into the GC. It is therefore important to consider if sample enrichment will be required in order to ensure that the amount of sample reaching the detector is sufficient for detection. The typical amount which can be detected using MS will vary and is dependent on the type and design of the mass spectrometer; this is discussed in later chapters. However, a typical single quadrupole MS operating in full scan mode requires around 1–100 pg to reach the detector from the column. If 1 µL is introduced into the GC then the concentration of this solution needs to be at least 1–100 pg per µL, which is equal to 1–100 ng mL^{-1} or 1–100 ppb. This assumes that the MS is working correctly, with regular checks being performed for MS tuning and calibration of the method (details of which are covered in later chapters).

Sample preparation can be performed:

- Manually, before the liquid extract is placed into an autosampler vial for manual injection or *via* a standard liquid autosampler for injection into the GC.

- Automatically with a special autosampler:
  - o For example, an XYZ robot can be programmed to perform each of the sample preparation steps which are usually performed manually.
  - o Or an autosampler designed to only perform the specific sampling or sample preparation technique, for example headspace analysis.

The main methods used to prepare a liquid sample for GC-MS analysis are:

- Using a liquid phase.
- Using a solid phase.
- Thermally, by heating the sample.

1.2.4 How Can I Prepare Liquid Samples with a Liquid Phase?

Liquid–liquid extraction is the mass-transfer of analyte(s) from one liquid phase into another liquid phase. The concentration of the analyte is based on the affinity of the analyte for the two immiscible liquids, usually water and an organic solvent.

1.2.4.1 *What Is Liquid–Liquid Extraction?*

The volatile and semi-volatile analytes present in a solution or a liquid sample can be enriched by extraction into another liquid. Enrichment can occur as it is possible to extract a large volume of the solution or liquid sample into a much lower volume of a second solvent. This can have the added benefit of removing any unwanted matrix (*e.g.* salts or proteins). This process of liquid–liquid extraction (LLE) has traditionally been conducted in a glass separating funnel. However, variations on this basic technique have been used in a wide variety of ways to optimise the enrichment ratio, including automating the entire process to GC-MS with the use of XYZ robots.

In traditional LLE the original solution (typically water) is extracted with an immiscible solvent (for example dichloromethane). The analytes of interest (typically non-polar in nature) have a greater solubility in the organic phase than the aqueous phase therefore partitioning of the analytes occurs whereby the majority of them move from the aqueous phase into the organic solvent. This process (when conducted in a separating funnel) is facilitated by shaking of the funnel to encourage mixing of the phases and movement of the analytes between the phases. The phases are then allowed to separate and the organic phase is decanted from the aqueous phase. The separating funnel facilitates this removal by a tap in the base of the funnel through which controlled quantities of the lower phase can be removed. The process is normally repeated at least three times with fresh organic solvent being used to partition the analyte completely into the organic portions.

This basic LLE process has been explored in many formats. Many of which aim to increase the enrichment process, make the partition process quicker or improve the extraction efficiency for analytes which have only a partial affinity for the organic/collection phase. For example, matrix modification is used to improve extraction and includes salting out and changing the pH, these are discussed further in this chapter under headspace analysis. This enrichment process has also been further refined and adapted to broaden its application and make use of smaller sample volumes to enable automation. A few of the more common adaptations are discussed below.

1.2.4.2   *What Is Dispersive Liquid–Liquid Micro-Extraction (DLLME)?*

In this adaption of the basic LLE, the liquid sample (typically water) has the extraction medium, typically a more dense solvent such as chloroform or dichloromethane (extraction solvent), rapidly introduced into the sample with a third solvent (the disperser) to aid the rapid and complete dispersion of the extraction solvent throughout the bulk of the sample solution. This is normally achieved by the use of a microliter syringe to introduce the solvents rapidly (less than 1 sec). The highly dispersed extraction solvent allows rapid partitioning of the analytes to occur between the liquid phases with potentially a higher efficiency than classical LLE owing to the very large surface area between the sample solvent and the extraction solvent. When the partitioning is complete, the extraction solvent is recovered by centrifugation of the sample separating the liquid phases. The extraction solvent phase is then removed and injected into the GC-MS for analysis (Figure 1.1). For this technique to be successful the precise ratio of the sample solvent, dispenser solvent and extraction solvent must be optimised.

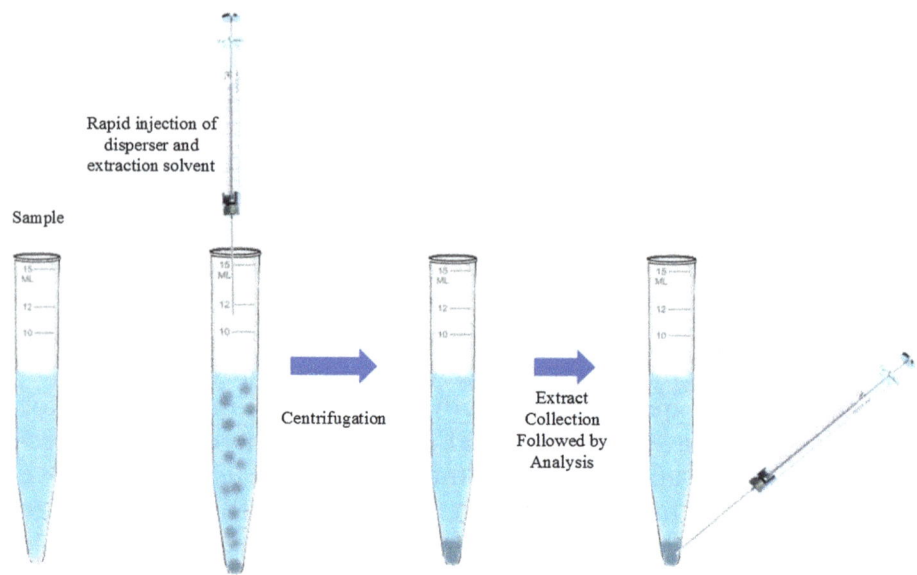

Rapid injection of
disperser and
extraction solvent

Sample

Centrifugation

Extract
Collection
Followed by
Analysis

Figure 1.1   Process Steps in DLLME.

1.2.4.3   What Is Single Drop Micro-Extraction (SDME)?

In this technique, an immiscible solvent droplet is held either in the sealed heated headspace above the sample or lowered into the bulk of the liquid sample from a microlitre syringe and is held suspended in this position whilst analytes from the sample partition into the solvent droplet. As with DLLME, the droplet presents a larger surface area to the sample and thus makes the liquid–liquid (or gas liquid partitioning in the case of headspace sampling) quicker than traditional LLE. It also operates at a much smaller scale. This approach uses microlitres of extraction solvent and 1–10 mL of sample. This means that the enrichment factors are high. That is, trace quantities of analytes in the sample are transferred into a few microlitres and injected onto the GC. Rather than these same quantities being transferred into a few millilitres of solvent then only a few microlitres being injected into the GC, with the bulk of the analytes not used in the analysis. There are a number of variables to control and optimise in this process such as the choice of extraction solvent, temperature of extraction and the extraction time.

1.2.5 How Can I Prepare Liquid Samples with a Solid Phase?

Solid phase extraction is a separation method that uses a solid phase to extract analyte(s) from a sample, based on their preferential affinity for the solid material over the sample.

1.2.5.1 What Is Solid Phase Extraction?

Solid Phase Extraction (SPE) and its related version dispersive SPE (dSPE), have long been used as extraction, clean-up and concentration processes and can perform all preparation tasks simultaneously. It uses solid sorbents to selectively remove either target analytes or unwanted interferences from samples prior to analysis. For example, contaminants can be eliminated through the following mechanisms:

- They are unretained and pass through during the loading step.
- Contaminants that are retained but are more polar than the analytes can be washed off.
- Contaminants that are less polar than the analytes can be left bound to the sorbent, for example, this is a good way of eliminating lipids from biological samples.

There are a number of steps involved in sample preparation using SPE, the typical principles include:

1. Sample pre-treatment: can include filtration to remove particles, changing pH to remove ionic interferences and dilution to minimise ionic and non-polar interferences.
2. Conditioning: clean the sorbent with 3–5 sorbent bed volumes of the analyte elution solvent.
3. Equilibrating: prepare the sorbent with 3–5 bed volumes of the sample solvent (known as solvation or wetting). Critical for a good efficiency and to obtaining the optimal sorbent capacity, otherwise this can result in break-through and poor recovery.
4. Loading: the sample is pushed or pulled through the sorbent, extracting and enriching analytes and some matrix components. A slow flow rate is often better, as too rapid a rate can lead to poor retention and recovery of analytes.
5. Air drying: can be used to remove residual water (optional).

6. Washing: analytes are retained and remaining interferents washed out. The solvent strength and pH must be optimised for selectivity, two half volume wash steps often work better than one entire volume.
7. Elution: analytes are eluted with a selective solvent and remaining interferents are retained. The solvent must be optimised and should be a stronger solvent than that used in the wash step and the elution volume should be determined. Large volumes can be evaporated (another step to optimise) ensuring volatile analytes are not lost, or a large volume injection into the GC may be performed.

Particulates will not pass through the SPE sorbent. Low levels will sit on top of the sorbent and can be extracted, but high levels should be pre-filtered before SPE if they make up a high percentage of the sample, otherwise they will block the cartridge.

There are a variety of SPE sorbent holders:

- Cartridges are used for the elution of small sample volumes (1–60 mL) with an optimal flow of 1–10 mL min^{-1}. They have a small surface area but a long path length and are suitable for low level particulates. There are multiple formats:
  - straight barrel;
  - large reservoir capacity;
  - Luer cartridges that can be run in series;
  - 96 well plate format;
  - micro-syringe format with an integrated SPE cartridge in the syringe needle (MEPS™).
- Columns are similar to cartridges, but can come in very large formats for preparative applications.
- Disks have a large surface area (47, 50 or 90 mm in diameter) and have faster optimal flows through the sorbent (200 mL min^{-1}) which are better for large volumes of sample (more than 100 mL) or samples with a higher proportion of particulates to be extracted.

There are a wide range of solid sorbents to choose from and the sample may be extracted through multiple sorbent beds, for example to selectively extract polar analytes and then non-polar analytes. The two main classes of sorbents are:

- Silica-based:
  - o Non-polar C18, C8 and Ph modified silicas: used for non-polar analytes, with retention through dispersion or hydrophobic/hydrophilic interactions; these analytes are easily adsorbed in a polar environment; they are eluted using a solvent of lower polarity.
  - o Polar unmodified silica (-OH groups), CN, NH_2 and –OH (diol): used for polar analytes with amino, hydroxyl, carbonyl, hetero-atoms, aromatic rings, double bonds and so forth, with retention through hydrogen bonding, dipole–dipole and π–π interactions; these analytes are easily adsorbed in a non-polar environment and they are eluted using a polar solvent.
  - o Ionic:
    - Cation exchange: silica modified with an acid (*e.g.* benzene sulphonic acid). Retention is through the interaction of charged, cationic groups on the analytes of interest and charged, anionic (negative) functional groups on the sorbent, *via* electrostatic (ionic) interactions. They are eluted with a solvent of high ionic strength, in which the presence of a large number of ions in the elution buffer creates competition with the analyte groups for the sorbent groups or through a pH change, in which either the charged analyte groups are neutralised, or the charged sorbent groups are neutralised, or through use of a buffer containing cationic species with a high affinity for the sorbent functional groups.
    - Anion exchange: silica modified with a quaternary amine, used for analytes with cationic groups: 1°, 2°, 3° and 4° amines and inorganic cations, and analytes with anionic groups: carboxylic and sulphonic acids, phosphates and so forth. The sorbent extracts analytes based on electrostatic interactions between the analyte of interest and the positively charged groups on the stationary phase. For ion exchange to occur, both the stationary phase and the sample must be at a pH in which both are charged. Ion exchange interactions are enhanced in low ionic strength samples and with counter ions with low selectivity (*e.g.* acetate, Na^+), they are eluted using a solvent of high ionic strength and high selectivity (*e.g.* citrate or Ca^{2+}).

- ■ Mixed mode columns: contain both ion exchange and hydrophobic ligands on a sorbent surface.
- • Polymer-based:
  - ○ Polymer or resin type: typically co-polymers with a single hydrophobic monomer plus a more polar monomer, can include a weak/strong ion exchange functionality, for example divinyl benzene resin with sulphonate groups. They have a higher loading (10–20% w/w) than silica-based sorbents (1–5% w/w).
  - ○ Molecularly imprinted polymers (MIPs): highly cross-linked polymer-based molecular recognition elements that are engineered to bind one target or a class of structurally related analytes. During synthesis a template molecule guides the formation of specific cavities in the polymer which are sterically and chemically complementary to the analytes with multiple non-covalent interactions. They are highly selective and use harsh wash conditions during sample preparation to obtain a lower background and therefore lower detection limits.

To develop an SPE method, there are a number of steps that must be followed:

1. Determine the structure and solubility of the analytes: it is necessary to determine the SPE sorbent to be used as this can help with solvent selection. Non-polar analytes have a Log P of more than 4.0; mid-polar analytes have a Log P of 1.5–4.0; and polar analytes have a Log P of less than 1.5. In reversed phase SPE (RP-SPE), the Log P value of the analyte must be greater than 1.5 in order to be retained on the non-polar sorbent. For ionisable analytes it is best to neutralise their charge prior to loading.
2. Select an appropriate SPE sorbent and weight: the capacity of the sorbent is the total amount of analyte that can be adsorbed under optimal conditions. This is 3–5% of the sorbent weight for silica sorbents and up to 30% for some polymers. Capacities should not be exceeded, otherwise the method will not be quantitative.
3. Determine the sample concentration and maximum loading.
4. Determine the optimal wash solvent: wash with as strong a solvent as possible that will not elute the analytes.
5. Determine the optimal elution solvent: should be optimised first using standards and no matrix interference. For example, for RP-SPE make up a range of elution solvents with 0–100% of organic

solvents in 10% steps and use these to elute each cartridge analysing and plotting the results.

- At 100% organic solvents the recovery should be 80–100%, if not a stronger elution solvent is required or a different sorbent.
- The lowest percentage of organic solvent with approximately 0% recovery is the best wash solvent.
- The lowest percentage of organic solvent with the highest recovery is the best elution solvent, leaving the less polar contaminants retained on the sorbent.

1.2.5.2   What Is Solid Phase Micro-Extraction (SPME)?

In some ways this technique is similar to SDME, however the liquid extraction medium is a high molecular weight liquid or a solid sorbent coated onto a fibre, both of which are similar to a GC column stationary phase. This is a solventless extraction technique. As with SDME, the fibre is exposed either to the headspace above the sample (headspace-SPME) or the liquid bulk (direct immersion-SPME) to extract and concentrate the analytes, as illustrated in Figure 1.2. Analytes from the liquid are partitioned or adsorbed from the sample onto the coated fibre. Once extraction is completed, the fibre is removed from the sample and either introduced into the injection port of the GC where the analytes are thermally desorbed or liquid desorption takes place with a solvent.

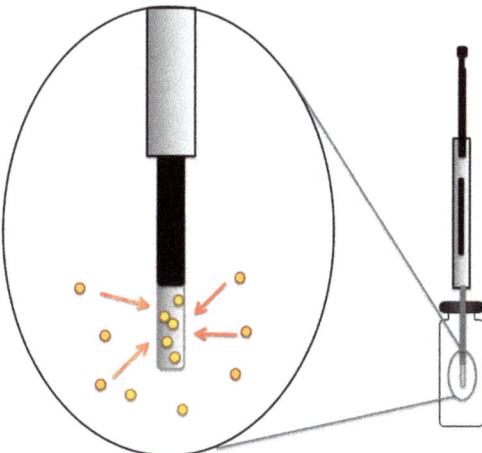

Figure 1.2   A schematic diagram of SPME.

The fibre is formed of 1–2 cm of fused silica coated with a stationary phase and bonded to a stainless-steel plunger. When piercing the vial or the GC inlet septum, the plunger is withdrawn within the needle to protect the fibre and then exposed to extract analytes or desorb the fibre. The phase type and thickness is chosen to match the characteristics of the analytes (selectivity). The amount of analyte adsorbed depends on the thickness of the phase and the partition coefficient of the analyte. As for GC columns, a thick phase is chosen for volatile analytes and a thin phase for semi-volatile analytes. There are two broad types of classification of SPME fibre phases:

- Homogeneous pure polymer coatings produce absorptive fibres in which the extraction occurs through partitioning, resulting in a good capacity and broad linearity:
  - Polydimethylsiloxane (PDMS):
    - 100 μm volatiles (MW 60–275)
    - 30 μm non-polar semi-volatiles (MW 80–500)
    - 7 μm mid- to non-polar semi-volatiles (MW 125–600)
  - Carbowax (PEG): 60 μm for alcohols & polar volatiles (MW 40–275)
  - Polyacrylate (PA): 85 μm for polar semi-volatiles (MW 80–300)
- Porous particles imbedded in a partially crosslinked polymeric phase producing both absorptive and adsorptive fibres which are better for extracting low concentration analytes:
  - PDMS/DVB: 65 μm for polar volatiles, amines and nitro-aromatics (MW 50–300)
  - Carboxen (CAR)/PDMS: 75/85 μm for trace-level volatiles (MW 30–225)
  - DVB/CAR/PDMS: 50/30 μm for flavour compounds C3–C20 (MW 40–275)
  - Activated carbon fibre (ACF): varying polarities, high temperature phase (380 °C).

The addition of a sorbent to a coating, for example strongly polar Carbowax PEG onto DVB polymer, increases the surface area and improves the extraction efficiency of polar molecules.

SPME analysis can be classified into two types, depending on where the fibre extracts the sample from:

- Headspace SPME (HS-SPME) extracts the analytes from the headspace above the sample and is therefore more sensitive for volatile analytes. The analytes must be predominantly in the headspace to be extracted, therefore matrix modification may be required to release the analytes into the headspace, this is discussed further later in this chapter concerning headspace analysis. HS-SPME equilibrates faster and keeps the fibre cleaner, minimising inter-ferents and prolongs fibre life.

- Direct immersion SPME (DI-SPME) inserts the fibre into the sample and therefore is more sensitive for analytes predominantly in the liquid sample. The sample must not damage the SPME fibre when extracting the analytes, as it is delicate and easily breakable.

Optimisation is again required to ensure that extraction from the sample into the fibre occurs in a reproducible manner. The method development steps are:

1. Select the most suitable SPME phases based on the chemistry of the analytes and phases:
   - Polarity: like separates like, therefore select a non-polar fibre for non-polar analytes and a polar fibre for polar analytes. Consider hydrogen-bonding, dipole–dipole, induced dipole, π–π and dispersion (non-polar) interactions.
   - Molecular weight: for example when selecting GC column stationary phases, choose a thicker phase for more volatile analytes (*e.g.* 100 μm) and a thinner phase for higher MW analytes (*e.g.* 7 μm).
2. Fibre conditioning: for the recommended temperature and time for the fibre phase chosen.
3. Sample preparation: for example, matrix modification for HS-SPME or sample dilution for DI-SPME for viscous samples.
4. Sample pre-incubation: heating and shaking temperature and time. In HS-SPME to establish an equilibrium between the sample and the headspace, the temperature must be optimised below the sample solvent boiling point. DI-SPME uses lower temperatures to minimise analytes in the headspace, but temperatures above room temperature to ensure reproducibility.
5. Extraction: the time the fibre is placed into the headspace or sample to extract the analytes, usually at the same temperature as the sample pre-incubation but with a gentle shaking at 100 rpm. The sample vial septum thickness should be rated for SPME, otherwise septum coring and fibre needle blockage may occur. The extraction time is generally longer for DI-SPME.
6. Desorption: the fibre is exposed in the GC inlet to quickly desorb analytes onto the column in 1–3 min. A narrow inlet liner with an internal diameter of 0.75 mm must be used in the GC inlet to obtain sharp peak shapes. Use of a pre-drilled inlet septum is recommended to prevent fibre breakage. The GC inlet temperature is

optimised to desorb the least volatile analytes but must be no higher than the recommended maximum operating temperature for the fibre, otherwise damage will occur.

7. Fibre bake-out: the fibre must be re-conditioned post-analysis to clean it for the next sample, it is recommended to bake-out the fibre at the fibre conditioning temperature, if a fibre-conditioning station is available.

There is a linear relationship between the initial concentration of the analyte in the sample and the amount ad/absorbed, therefore the technique is quantitative, however, it is highly recommended to use an internal standard (IS) to improve the accuracy. The amount of analyte extracted is not related to the sample volume, therefore the technique can be used for field sampling of lakes, air, and so forth. However, for high concentration samples the sample volume should be no greater than 5 mL if calibrating, as the amount of analyte removed by the fibre is not sufficient to change the sample concentration.

1.2.5.3 What Is Dispersive Solid Phase Extraction and QuEChERS?

The QuEChERS method has gained popularity for its wide applicability in the preparation of samples being analysed for pesticide residues in food. It has been adopted in a number of standards for analysing pesticide residues including EN15662 and the AOAC Official method 2007.01 as an alternative to more traditional LLE. Its key advantage is a reduction in solvent use and ease of automation, which reduces costs overall. It is a standardised approach, made up of three steps:

1. Sample preparation: the sample is ground or homogenised and an IS is added.
2. Sample extract clean-up: sample extraction and clean-up vary somewhat depending on the sample and target analytes however, the methodology is defined by the addition of an extraction solvent, buffering system (such as sodium acetate or acetic acid) and drying agent (typically magnesium sulphate). The organic layer is then cleaned-up further by dSPE using a primary secondary amine (PSA) or a C18 (octadecyl) modified silica-based or graphitized carbon black (GCB) sorbent.
3. Sample analysis: centrifugation is used to separate the solids from the liquid extracts, which are then injected into a GC-MS.

1.2.6   How Can I Prepare Liquid Samples Thermally?

Heat can be applied directly to a liquid-phase sample to release the volatile analytes for analysis. The main techniques used for liquid samples, in order of increasing sensitivity, are static headspace, dynamic headspace and purge-and-trap analysis.

1.2.6.1   What Is Static Headspace (SHS) Analysis?

Headspace (HS) can be defined as the gas space above a sample when it is placed in a chromatography vial and therefore it is the analysis of the analytes present within that gas. Samples that can be analysed using HS analysis include anything that can fit in the vial and release volatile compounds, these include liquids and solids; here we will focus on liquid samples. Usually, heat is applied to the vial to assist the release of the compounds. How much heat can be applied depends on the boiling point of any solvents in the sample and the maximum temperature of the instrument and the vial. For most HS analyses this is usually in the thermal desorption region, below 350 °C, in which carbon–carbon bonds are not broken.

The sample is placed into a HS screw-top or crimp-top vial, ranging in size from 2–22 mL depending on the HS autosampler, with 10 or 20 mL vials being typically used. The size is important when considering the phase ratio, as detailed below. The HS vial containing the sample is then sealed with a septum and cap. The septum is usually silicone-PTFE, but can also be rubber, it should be selected to ensure that it:

- Provides a gas-tight seal;
- Does not absorb the analytes from the sample;
- Does not add bleed compounds to the vial HS.

PTFE (polytetrafluoroethylene) is very inert and should be facing the sample, however once pierced there is a route to the silicone which can bleed or absorb analytes. If possible, it is best to not pierce the septum when adding to the vial, for example when adding matrix modification reagents or spiking with an IS, these should be added before capping. Using the wrong septum for an application can have a large effect on the reproducibility and sensitivity. The vial, cap and septum should all be rated to the temperatures used in the method, especially if reactions, such as derivatisation, will take place in the vial. In which case, it is recommended to use a 'venting' septum, which has a weakness in case of too much vial pressure as it is preferable for the septum to break rather than the vial!

Once the sample, any standards and matrix modification reagents are sealed in the HS vial, it is then usually heated and shaken. The heating of the system causes the equilibrium of volatile analytes within the liquid to shift towards the gas phase in accordance with their relative affinity for the liquid and gas phases, shaking reduces the time taken to achieve this. The equilibrium is dependent on a number of factors which are beyond the scope of this chapter, but are referenced at the end under further reading. In simple terms, the most volatile analytes will partition towards the gas phase and the least volatile will remain in the liquid phase. The concentration of the analytes in the headspace will be affected by a number of factors including the pressure, ionic strength (in the liquid phase) and temperature, plus the length of time required for an equilibrium to be established. Matrix modification, for example the addition of salt, a change of pH, addition of a co-solvent or derivatisation can help particular analytes to move into the gas phase, for example salt is useful for enabling polar analytes to migrate out of a polar liquid matrix. This partition process is a dynamic equilibrium. At a fixed temperature, the partition coefficient is defined as the ratio of a given analyte in the gas and liquid phases. The ratio between the volume of the sample and volume of HS (phase ratio) has an impact on the recovery of those analytes with good partition coefficients, with larger sample volumes and smaller HS volumes giving this technique better sensitivity. Analytically, the aim is to migrate the volatile analytes of interest as much as possible into the gas phase whilst leaving behind the potentially interfering non-volatile sample constituents. To do this the equilibration time, temperature, matrix modification and phase ratio all need to be optimised to produce a sensitive, reproducible and robust method, the use of an IS is always recommended.

The HS can then be sampled using a gas tight syringe or heated valve and transfer line to transfer the analytes out of the HS vial and into the GC for analysis. Sampling of the HS upsets the equilibrium and each sample taken from the HS vial will result in an overall reduction in the analyte concentration.

1.2.6.2   *What Is Dynamic Headspace Analysis?*

A variation on this theme is the concept of dynamic headspace (DHS) analysis. DHS takes place in a purged vial. A flow of carrier gas is purged over the sample and continuously transfers the volatile analytes emerging from the sample into a trap, in which they are concentrated. The equilibrium is never reached and after a fixed period the purging stops and the trap is thermally desorbed, to transfer the analytes into the GC column for analysis. Alternatively, it can be eluted with an organic solvent for analysis using different techniques. As most of the analyte molecules are recovered, DHS is more sensitive than static HS, but only for those analytes with low partition coefficients. The sample is treated in the same way as for static HS, with heating of the sample vial and matrix modification to release the analytes from the sample matrix. This approach requires a further step to be optimised. Namely, the selection of the optimal trap adsorbent(s) that both traps and releases the analytes quantitatively with no break-through, irreversible adsorption or catalytic breakdown, while providing the best recovery of the analytes. It is also advantageous to choose a selective adsorbent that does not trap the matrix, for example a hydrophobic adsorbent for aqueous samples (more detail on sorbent selection is given under thermal desorption in this chapter). The optimal adsorption and desorption temperatures and the flows must also be determined and care must be taken to minimise activity, dead volumes and cold spots through the more complicated flow path and valve.

1.2.6.3 Purge-and-trap Analysis

Purge-and-trap (P&T) analysis is even more sensitive down to ppt levels. It is very similar to DHS, however rather than the gas flowing over the top of the sample to remove the HS, the inert gas bubbles through the liquid sample to actively remove analyte molecules and sweep them into the trap. Care must be taken to prevent foaming of the samples which would result in contamination of the system and the trap. In newer instrumentation foam sensors are used and an anti-foaming agent can be added to reduce the likelihood of occurrence.

1.3 How Do I Sample and Prepare a Solid for GC or GC–MS Analysis?

It is important to differentiate between the need to analyse a solid sample (which is a mixture or a single substance) and the need to extract the volatile and semi-volatile analytes contained within a solid sample.

Some solid-phase samples must be sampled *in situ*, for example volatiles emitted from building materials, car interiors or plants. For

others, a sub-sample can be taken back to the lab and a portion of
this analysed, for example soil, drugs, washing powders, meteorites
and food.

Solid samples can be prepared in a similar fashion to liquid samples. Care should be taken to select a solvent which dissolves all of the
sample as rapidly as possible. As with liquid samples, the choice of
a suitable solvent is very much dependent on the sample properties.
A solvent should be chosen which dissolves the sample, is amenable
to gas chromatography (see Section 1.2.2) and in which the analyte
is stable for the time required for the analysis. It may be necessary
to transform the analyte into a more volatile substance to make GC
analysis possible. Further details on derivatisation (Section 1.4) and
pyrolysis (Section 1.3.2) are provided later in the chapter.

As with liquids, storage prior to analysis under controlled conditions is important to ensure the sample presented for analysis is representative. This may require the solid sample to be protected from
light, water ingress and extremes of temperature, which along with
other factors may alter the solid prior to analysis. Particular care must
be taken with biological samples.

Treatment of the solid to maximise solubility in the chosen solvent may include milling the sample to decrease particle size and
to increase the surface area, or use of sonication to encourage rapid
solubilisation.

Solid samples, for which there is a need to analyse the volatile analytes they contain but for which no suitable solvents exist, can be
prepared using other treatments including: HS analysis (see Section
1.2.6), TD and pyrolysis.

1.3.1   Can I Use Thermal Desorption?

Thermal desorption is described for the concentration and transfer of
gas phase samples to the GC-MS for analysis in Section 1.1.3.1. Small
solid or viscous liquid samples can be directly thermally desorbed, at
temperatures up to 350 °C so that no chemical bonds are broken, by
placing them in a conditioned TD tube. Larger solid samples can be
placed in a micro-chamber or macro-chamber up to 1 m^3, in which
they are heated and an inert gas sweeps the analytes into a packed TD
tube. A schematic of the process is shown in Figure 1.3.

The parameters that need to be optimised are:

- Sample amount (weight) and size: large differences in the surface
  areas for the same sample weight will affect the time taken to
  fully desorb the sample.

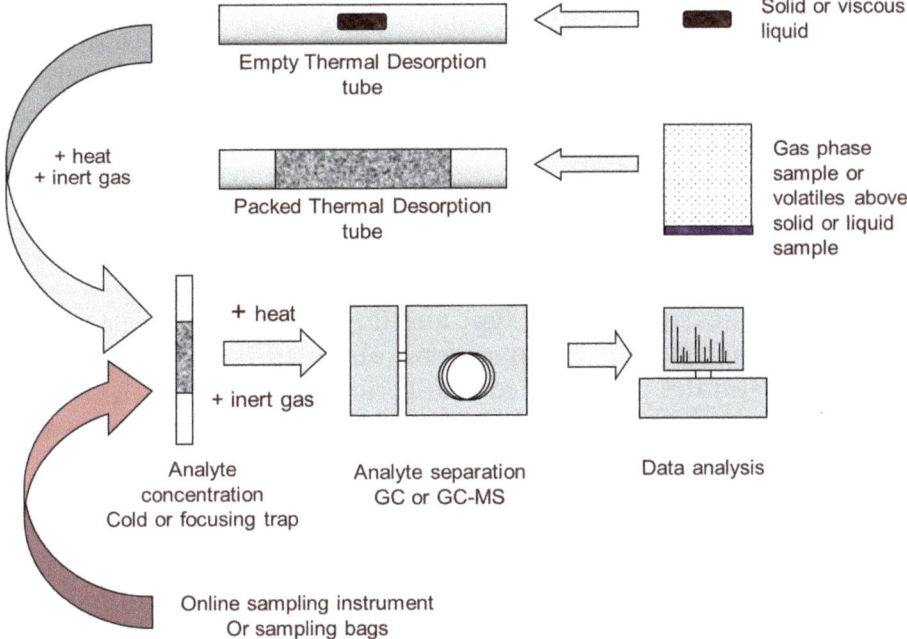

Figure 1.3 Process of thermal desorption. Courtesy of Anthias Consulting Ltd.

- Dry purge: to remove water (and air) as water will produce a large vapour volume leading to contamination of the system.
- Pre-purge: the TD tube must be purged with carrier gas to remove oxygen and prevent oxidation of the sample.
- Desorption temperature: hot enough to extract the analytes but not too hot that the matrix carbonises and absorbs the analytes or causes reactions.
- Desorption time: the full sample needs to be heated and all volatiles extracted for reproducibility.
- Cold trap: the sorbent material used (see Section 1.1.3.3), trapping and desorption temperatures and time.
- Split flows: from the TD tube to the cold trap, then from the cold trap to the GC column. For low concentrations, a small split from the cold trap to the GC column is always advisable to produce better peak shapes, rather than using a splitless transfer.

1.3.2 When Should I Use Analytical Pyrolysis for Solid Samples?

Pyrolysis can be used to gain information about solid samples through thermal decomposition of the sample in a controlled environment (usually inert). The sample is rapidly heated to temperatures above

350 °C, usually in the region of 600–1400 °C, in which chemical bonds are cleaved within the macromolecular structure producing low molecular weight, more volatile analytes that are specific units of that macromolecule. These are volatile and can be transferred to the GC-MS for analysis, either through a heated transfer line or directly from a pyrolyser on top of the GC inlet. The collection of analytes can give an insight into the sample. Pyrolysis chromatograms, called pyrograms, are typically complex with many individual analytes obtained, the result produced is dependent on the pyrolysis temperature, heating rate, gas type and flow used for the inert atmosphere and the sample. The pyrogram obtained is a fingerprint of the sample and the profile or identification of the key analytes can be invaluable for particular applications such as the forensic analysis of paints or pigments or other high molecular weight analytes for which direct analysis is difficult, such as plastics, synthetic polymers, rubber, rocks, oil or wood. The pyrolysis products of polymers can give an insight into the composition of the polymer. In some cases, the polymer may simply depolymerise to yield the monomer or monomers. In other cases, this process may yield more complex products.

The mechanism of rapid heating varies depending on the type of pyrolyser. Common types include: inductive heating, resistive heating, a furnace, laser and heating within non-specialist pyrolysers such as within a programmable temperature vaporiser (PTV) inlet.

All pyrolysers share a common aim, to rapidly heat (20–1500 °C ms^{-1}) to a high temperature 600–1400 °C, with a short temperature rise time (TRt) and to hold for a short pyrolysis time (2–60 s), sufficient to fully pyrolyse the sample, but not produce secondary reactions. An example of a temperature-time profile (TTP) is shown in Figure 1.4a. Pyrolysis methods include:

- Isothermal: analysis at a single temperature with a single pyrogram produced. Includes flash pyrolysis with a TRt of approximately 8 ms.
- Sequential: the sample is heated to the same temperature multiple times with a pyrogram produced for each pyrolysis. This method is useful for kinetic studies to understand degradation and formation rates and activation energies. It is also used to determine if different substances give the same pyrolysis products at different rates.

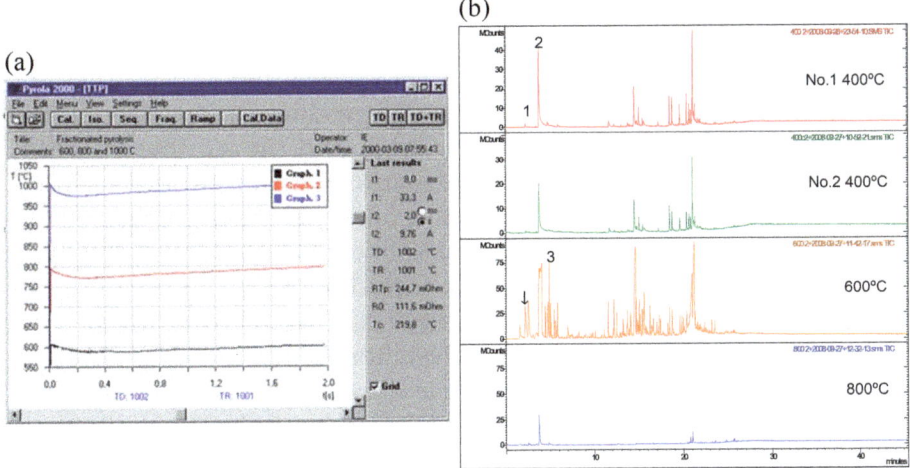

Figure 1.4 Example of fractionated pyrolysis of toner: (a) the temperature-time profile (TTP); (b) pyrolysis number 1 @ 400 °C (red), pyrolysis 2 @400 °C (green), pyrolysis 3 @600 °C (orange) and pyrolysis 4 @800 °C (blue). Courtesy of Pyrol AB, Sweden.

- Fractionated: the sample is heated to different increasing temperatures with a pyrogram produced at each temperature. Used in complex samples and for thermally labile analytes. An example of fractionated pyrolysis of toner at three different temperatures is shown in Figure 1.4b, along with their TTP in 1.4a.
- Pyrotomy: the sample is pyrolysed for a very short time (20 ms) to only pyrolyse the surface of the sample in contact with the filament. Used for non-homogenous samples in which the sample can be analysed layer by layer.

Sample preparation is very straight forward, microgram to milligram quantities (pyrolyser dependent) of a sample are placed onto a ribbon filament or into a quartz tube or metal cup which is then sealed into the pyrolyser and rapidly heated or dropped into a furnace. Sub-pyrolysis temperatures may also be employed to thermally desorb any volatile species which may be present within the matrix, such as plasticisers or other additives. Pyrolysis parameters that need to be considered are:

- Sample size: a small sample size, which has a constant weight and shape, is essential to ensure that all of it degrades fully and rapidly otherwise it may result in poor reproducibility.

Samples are rarely homogenous, therefore quantitation can be difficult.

- Pyrolysis temperature: depends on the sample and problem to be solved. Higher temperatures can produce larger, analytically more significant molecules. Too high and the sample degrades producing uncharacteristic information with simultaneous bond breakages to create very small and non-specific free radicals and molecules, carbonisation can also occur. Too low, and the sample may not be totally pyrolysed. Different samples have different degradation rates and can produce exothermic or endothermic reactions resulting in the pyrolysis temperature being different to that set in the unit. To optimise the pyrolysis temperature, the analyte area should be plotted against the pyrolysis temperature.
- Pyrolysis time and ramp rate: long pyrolysis times, including slow pyrolysis ramp rates, can produce secondary reactions in some samples, resulting in analytes being identified that were not originally part of the sample structure.

Pyrolysers can also be used for derivatisation, for example thermochemolysis (thermally assisted hydrolysis and methylation) is used to selectively break ester and ether bonds, before the products are methylated and analysed. The sample can then be further analysed by pyrolysis.

The main disadvantage of this technique is that often the chromatograms are very complex and contain many analytes to identify, making it challenging to reconcile the information obtained with the sample. However, if time is taken to construct reference libraries and deconvolute the complex GC-MS dataset this technique can provide valuable information.

1.3.3   What Liquid Extraction Techniques Can Be Used?

There are a wide number of liquid extraction techniques for the preparation of solid-phase samples, the more common ones include reflux, soxhlet and static extraction, either in an open or sealed vessel.

Each of these techniques relies on the action of a solvent on a solid sample, the solid matrix itself is insoluble in the selected solvent but the constituent volatile and semi-volatile analytes are soluble and can be extracted by the action of the solvent penetrating the solid matrix over time.

1.3.3.1 What Is Static Extraction?

This is perhaps the simplest form of liquid extraction. Here, the solid sample is simply placed into the liquid solvent and left for a fixed period of time to allow the solvent to penetrate the solid and extract the volatile and semi-volatile constituents through diffusion. To aid this process, and indeed any of the described extractions from a solid sample, it is possible to increase extraction efficiency by:

1. Increasing the surface area of the solid undergoing extraction. This is typically done by grinding or cutting the solid into smaller particles or parts. If volatiles may be lost, cryogenic grinding (also known as freezer grinding or milling) can be employed using liquid carbon dioxide or liquid nitrogen.
2. Increasing the temperature of extraction, this increases the diffusion rate of the volatile and semi-volatile analytes from the solid into the solvent.
3. Increasing the time for which the solid and solvent are in contact with each other. Diffusion can be a slow process and will become slower as the equilibrium is reached between the concentration of the analyte in the solid and the solvent. It may be necessary to repeat the extraction of the solid with fresh solvent to remove all of the analytes.

Variations of this also include the use of sonication (sound energy) from acoustic cavitation owing to the formation, growth and implosive collapse of bubbles in a liquid, this can be used to speed up the extraction process in several ways. The samples are normally placed in a laboratory ultrasonic bath (filled with water), the extraction solvent is placed in a beaker or a round bottom flask which is secured within the ultrasonic bath. A high frequency transducer in the kHz range causes cavitation of the water in the bath and the extraction solvent. This energy speeds up the extraction both through agitation and heating of the extraction solvent. The disadvantage of this method is that the energy levels can be hard to control and therefore the robustness and reproducibility can be poor if the energy levels vary.

1.3.4   What Is Microwave Extraction?

Another variant is the use of microwave energy to heat the extraction solvent in a sealed container above its boiling point. The effect of elevated pressures and temperatures can speed up the extraction of analytes. Different solvents will be heated by the microwave energy more or less, depending on their dipole moment. The solvent dielectric constant can be used to select extraction solvents that interact with the microwave energy the most and thus are quickly heated. Generally, the more polar the solvent the more efficiently it heats, thus water is heated very quickly and hexane not so quickly.

Microwave extraction can be influenced by a number of factors which are common to the other extraction methods described here such as the extraction time, temperature, microwave power and the solid sample characteristics, however, the solvent choice is the most important. Solvent mixtures can be used to optimise the characteristics of the extraction including the addition of salts and hexane/water or acetone/hexane mixtures to improve the extraction efficiency and ensure complete extraction. The hexane swells the solid matrix and solubilises the non-polar analytes, the acetone interacts with the microwave energy to heat the extract. Clearly some quite complex optimisations are required once mixtures are introduced, but this makes microwave extraction very flexible. Microwave extraction units also allow precise monitoring of microwave energy levels, temperatures and pressures within the sealed unit containing the extraction solvent and sample, leading to much greater control over the extraction conditions.

1.3.4.1   What Is Reflux Extraction?

In reflux extraction, the solid sample is placed into a round-bottom flask, the solvent is added and a refluxing condenser is placed on top. The flask is heated to above the boiling point of the solvent, the solvent boils and is returned to the round bottom flask *via* the condenser. Over time this refluxing action removes and collects the volatile and semi-volatile analytes. After a fixed period of time the solvent is allowed to return to room temperature and is then available to be analysed as a liquid extract using GC-MS. Extraction times can be quite long, with periods of 16 hours or more not being uncommon, depending on the nature of the solid, the analytes and how complete and reproducible an extraction is required. It is difficult to process large number of samples, with long extraction times plus the glassware, heat source and condenser set required for each sample limits the number of samples that can be processed in parallel. The volume of solvent used is also quite high, with associated safety aspects, costs and waste disposal to consider. The supply of water may also be an issue and all these factors contribute to the environmental impact of analysis.

1.3.4.2 What Is Soxhlet Extraction?

Soxhlet extraction is a similar process to reflux extraction with the difference being that the solid sample is suspended above the solvent in a round bottom flask, held within a specialised type of condenser, the soxhlet condenser. Therefore, the solid sample is only exposed to the condensing solvent after it has evaporated from the round bottom flask, rather than the boiling solvent. This means that the temperature of extraction is slightly lower and the solid is always exposed to just boiled pure solvent without the extracted material, which collects in the flask. The technique potentially improves the extraction efficiency as the extraction media is never saturated by the extractants. Soxhlet extraction shares many of the same characteristics of reflux extraction, it is simple and inexpensive to setup and is an effective extraction technique if time is given for extraction to occur. Its disadvantages are the speed of extraction and a limited ability to process large volumes of samples in a timely manner. It also uses relatively large volumes of solvent, which adds complexity if the concentration of the analytes in the solid are low and are therefore further diluted by the extraction procedure. Therefore, solvent concentration may be required, see Section 1.5.

1.3.5 What Is Pressurised Fluid Extraction?

Pressurised fluid extraction (PFE) is a generic term for a technique which uses high temperatures and pressure to reduce the extraction time. It includes subcritical and supercritical fluid extraction.

Accelerated solvent extraction (ASE) is a form of PFE and is a registered trademark of Dionex. PFE extraction systems use temperatures of up to 200 °C and pressures of around 1500 psi within an extraction cell. The extraction cell contains the solid sample and the solvents are introduced either using static extraction or a dynamic extraction step and often a combination of the two. The temperature and pressure reduce the extraction time and reduce the volume of solvent needed to extract the solid. As with microwave extraction the extraction is more controlled owing to precise monitoring of the temperature and pressure, but unlike microwave extraction, it is possible to dynamically adjust the extraction solvent to optimise the extraction process. There are numerous parameters which affect the extraction making it necessary to undertake careful optimisation to achieve a reliable and effective result, when properly optimised it is a very effective technique!

Supercritical fluid extraction (SFE) is a sub-category of ASE. This extraction technique uses the physical properties of supercritical fluids to improve the extraction. A supercritical fluid is a substance above its critical point. A fluid above its critical point has properties which enhance the extraction of solids such as enhanced effusion into a solid (gas-like) and the ability to dissolve solids (liquid-like). A substance is brought above its critical point when pressure and temperature are applied. A common fluid is supercritical carbon dioxide because its critical point is within a convenient pressure (74 Bar), and temperature range (31 °C) and it is easily available (Figure 1.5). Many other fluids can be used but they have practical issues relating to the toxicity and availability. Carbon dioxide is frequently modified during the extraction process by the addition of co-solvents such as methanol. In SFE, the fluid is pumped through an extraction cell containing the solid to be extracted. As with microwave extraction and ASE the evaluated pressure in a sealed container enhances the extraction reducing the extraction time.

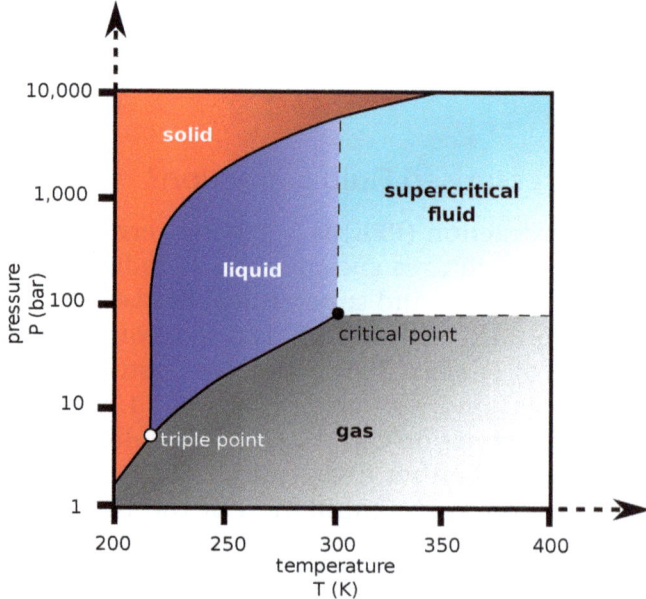

Figure 1.5   The critical point of carbon dioxide.

Table 1.1 Comparison of extraction techniques for solid samples.

	Speed	Cost	Complexity of development	Effectiveness	Reproducibility	Ease of automation
Reflux	Slow	Low – glassware and solvents	Low	Variable[a]	Variable[a]	Poor
Soxhlet	Slow	Low – glassware and solvents	Low	Variable[a]	Variable[a]	Poor
Static	Slow	Low – glassware and solvents	Low	Variable[a]	Variable[a]	Poor
Sonication	Slow/medium	Ultrasonic bath cost	Low	Good	Poor[b]	Poor
Microwave	Quick	High	Medium	Good[c]	Good[c]	Good
ASE	Quick	High	High	Good[c]	Good[c]	Good
SFE	Quick	High	High	Good[c]	Good[c]	Good

[a]Can suffer from low and variable recoveries owing to saturation of the solvent, the time taken to complete extraction is long therefore incomplete extraction can occur if extraction times are short.
[b]Very dependent on the control of the sonication energy.
[c]If method development is successfully undertaken these methods can be highly effective and reproducible.

As you might expect each extraction technique has its strengths and weaknesses these are summarised in Table 1.1. Which technique is selected will depend on the application as well as these factors.

1.4 When Should I Use Derivatisation in GC–MS Analysis?

Derivatisation is used for various reasons in GC-MS analysis and these can be sub-divided into four main groups:

1. To make the analytes more volatile, and thus amenable to gas chromatography.
2. To remove functional groups which make gas chromatography problematic, for example hydroxyl, carboxyl or amine groups.
3. To improve detectability, this may overlap with point 2 to a degree if functional groups are changed to improve the chromatography. Furthermore, alternative functional groups may be formed which generate a better response in the detector.
4. To improve the thermal stability of the analyte to facilitate GC-MS analysis.

The general approach is to chemically transform the target analyte(s) through the use of a reaction with a derivatisation reagent either prior to the subsequent analysis as a separate sample preparation step or in the injection port of the gas chromatograph. PTV injectors are useful for this. There are four commonly employed derivatisation modifications: these are silylation, acylation, alkylation and esterification.

There are many different reagents which have been developed and these have been reviewed and discussed elsewhere.

Silylation aims to replace an active hydrogen on an OH, SH or NH group in a molecule (Scheme 1.1). This has the effect of increasing the volatility and reducing the interaction with the GC flow path residual silanol groups, thereby reducing peak tailing.

$$\text{Sample-OH} + \text{R}_3\text{Si-X} \longrightarrow \text{Sample-O-SiR}_3 + \text{HX}$$

Scheme 1.1

The ease of silylation follows predictable trends as per nucleophillic substitution and the basic nature of the leaving group on the silylating agent (Table 1.2)

Table 1.2   Common silylation reagents.

Reagent	Abbreviation	Applications
N,O-bis(trimethylsilyl) acetamide	BSA	OH, COOH, amides, amines
N,O-bis(trimethylsilyl) trifluoroacetamide	BSTFA	OH, Ar–OH, COOH, carbohydrates, amides, amines, acid anhydrides, sulphonamides
Dimethyldichlorosilane	DMDCS	Deactivating glass
Hexamethyldisilazane	HMDS	OH, Ar–OH, COOH, amines
N-t-butyldimethylsilylimidazole	TBDMSIM	Unhindered OH and Ar–OH
Trimethylchlorosilane	TMCS	Silylation catalyst; used with other reagents
N-trimethylsilylimidazole	TMSI	OH, COOH, carbohydrates, fatty acids, sulphonic acids, Ar–OH, R–SH
BSA + TMCS		OH, alkaloids, amines, biogenic amines, carbohydrates, COOH, Ar–OH, steroids
BSA + TMCS + TMSI		OH, amines, amides, amino acids, COOH, Ar–OH, steroids
BSTFA + TMCS		OH, alkaloids, amides, amines, biogenic amines, COOH, Ar–OH, steroids
HMDS + TMCS		Amino acids, amipicillin, carbohydrates
HMDS + TMCS + pyridine		OH, bile acids, carbohydrates, Ar–OH, steroids, sterols, sugars
TMSI + pyridine		C=O, steroids

Acylation replaces an active hydrogen with an acyl group (Scheme 1.2). This increases volatility and can also improve thermal stability and reduce interaction with the active sites through the GC (Table 1.3)

$$\text{Sample-OH} + \text{R-}\overset{\overset{\displaystyle O}{\|}}{\text{C}}\text{-X} \longrightarrow \text{Sample-O-}\overset{\overset{\displaystyle O}{\|}}{\text{C}}\text{-R} + \text{HX}$$

Scheme 1.2

Esterification is a subset of alkylation in which organic acids are reacted with alcohols to produce the corresponding ester. A common example of this is fatty acid analysis owing to their corresponding methyl esters (FAME analysis). Transesterification is used to displace one ester group with another to produce a new ester. This is commonly deployed in the analysis of fats and triglycerides (Table 1.4).

Table 1.3   Common acylation reagents.

Reagent	Abbreviation	Applications
Acetic anhydride		OH, Ar–OH, carbohydrates, amines
Trifluoroacetic acid	TFA	Amides, amines, C=O, OH, sulphonamides, silyl catalyst
Trifluoroacetic acid	TFAA	OH, amino acids, amides, amines, Ar–OH, steroids
Pentafluoropropionic acid anhydride	PFPA	OH, amino acids, amides, amines, Ar–OH, steroids
Heptafluorobutyric acid anhydride	HFPA	OH, amino acids, amides, amines, Ar–OH, steroids

Table 1.4   Common alkylation/esterification reagents.

Reagent	Abbreviation	Applications
Boron trichloride-2-chloroethanol		Esterifying/halogenation for ECD work, phenoxy acid herbicides
Boron trichloride-Methanol	BCl_3–MeOH	COOH, transesterification
Boron trifluoride-Butanol	BF_3–BuOH	Short chain carboxylic acids, transesterification
Boron trifluoride-Methanol	BF_3–MeOH	Long chain carboxylic acids, transesterification
Methanolic sulphuric acid	MeOH–H_2SO_4	COOH, transesterification
Methanolic base (metallic sodium in methanol)	Na in MeOH	Transesterification of triglycerides, cholesteryl esters, phospholipids
Methanolic HCl	MeOH–HCl	Fatty acids
Pentafluorobenzyl bromide	PFBBr	Halogenated derivatives of COOH, mercaptans, Ar–OH, sulphonamides
Trimethylanilinium hydroxide	TMAH	Carbamates, hydroxyl amines, barbiturates

1.5 How Do I Know How Well My Sample Preparation Has Worked?

Sample preparation has the advantages of removing the matrix or concentrating the analytes, however this process can also cause the loss of analyte molecules. Analyte recovery is an important parameter to check when developing and validating a method. Sample preparation is the largest source of error in the analysis of a sample by GC-MS, therefore it is important to optimise the entire GC-MS method, including sample collection and sample preparation. The recovery criteria often dictates that it should be more than 80%, however if the accuracy and precision of the method are good, then a lower percentage recovery may be acceptable. Reproducible recovery indicates that the method is robust.

Recovery can be calculated by:

$$\%\text{Recovery} = \left(\frac{Area\,A}{Area\,B} \right) \times 100\%$$

In which *Area A* is equal to the peak area of the prepared sample (which could be an extraction) and *Area B* is the peak area of the prepared (extracted) sample matrix in which the analyte was added after the sample preparation step (post extraction).

2 How Do I Introduce My Samples into the GC Column?

2.1 How Do I Inject My Sample into the GC?

In order to successfully analyse a liquid sample using gas chromatography (GC) it must be transformed into the gaseous state. This is normally achieved by introducing a small portion of the liquid sample into the injection port (also known as inlet) of the gas chromatograph (or in the case of on-column GC, the column) in which it is heated above the temperature required to turn it into a vapour and chromatographed in a gas. The optimisation of the process, by which the liquid sample is placed into the GC column, is one of the key steps used to ensure a successful analysis. This is a two-part process:

1. Successfully inject the sample into the GC inlet.
2. Successfully transfer the analytes into the GC analytical column.

With liquid samples, it might only be necessary to ensure that the sample injected into the GC is suitable for accurate determination of the analytes that you wish to measure. This means ensuring that the components you want to analyse are within the working range of both the GC and associated mass spectrometer (MS). Although the working range of a GC-MS is quite broad, dilution or concentration of the sample constituents may be necessary.

Gas Chromatography-Mass Spectrometry: How Do I Get the Best Results?
By Diane Turner, Mathias Schäfer, Steven Lancaster, Imran Janmohamed, Anthony Gachanja and Jason Creasey
© Diane C. Turner, Mathias Schäfer, Steven Lancaster, Imran Janmohamed, Anthony Gachanja and Jason Creasey 2020
Published by the Royal Society of Chemistry, www.rsc.org

The GC-amenable sample may be introduced into the GC using a manual syringe, autosampler or valve. In the early days of capillary column GC, it was found that it was necessary to use injection systems in which the solvent and sample are volatilised before entering the column. During volatilisation techniques, the sample is introduced into the heated inlet using a micro-syringe. Gas-tight syringes are available for the injection of gases and vapours. The accuracy of the quantitative injection of the sample into the column depends on:

- the rate of sample introduction;
- the syringe dead volume;
- the heating rate of the syringe needle by the inlet;
- the method used for handling the syringe during sample introduction.

The most common methods for liquid sample introduction using a syringe with injection into the GC inlet are:

1. Filled needle, in which the sample is taken only into the needle. Injection is accomplished by placing the needle into the injection zone and through evaporation the sample is transferred into the inlet without any mechanical movement of the plunger.
2. Cold needle, in which the sample is drawn into the syringe barrel followed by an air-plug, the syringe needle is empty and is inserted into the heated injection zone. The sample is injected immediately by depressing the plunger and the remaining sample in the needle is allowed to evaporate before withdrawal through a post-injection delay.
3. Hot needle, this is the same as cold needle but upon inserting the needle into the injection zone, the needle is allowed 3–5 seconds to heat up through a pre-injection delay, before depressing the plunger so that evaporation takes places from inside the needle.
4. Solvent flush, in which a solvent plug is drawn into the syringe barrel before the sample. The solvent and sample may or may not be separated by an air plug. Injection is performed as described in the cold needle technique above.
5. Air flush, the injection is performed in the same way as the solvent flush described above, except that an air plug is used instead of a solvent plug.

The injection may be performed manually, although autosamplers are now more common and are programmed to do any of the above modes of injection. Typically, for capillary column GC, the values for injection volumes are 0.1–3.0 µL. Poor injection precision may result from needle dead volumes, the sample adhering to the needle surface and back flushing of the sample past the plunger. Sample mass discrimination is a problem usually encountered when a sample containing components that have a wide boiling point range are injected into the hot vaporising inlet. When the sample and solvent are injected, some will be left behind in the syringe needle as the more volatile sample components and the solvent will distil from the syringe needle faster than the less volatile components, this is known as syringe mass discrimination. This will mean that the sample reaching the column will be of a different composition from the original sample; it will contain a higher proportion of the more volatile sample components. The hot needle and solvent flush techniques described above are preferred for minimising sample discrimination during injection. Use of a post-injection delay of 0.5–2.0 s after the plunger has been pushed down as rapidly as possible, with no pre-injection delay is also a preferred method and is very effective.

Another problem is adsorption or catalytic decomposition of labile substances by the needle when using vaporising inlets. To minimise this problem, deactivated fused silica syringe needles and cold on-column or PTV injection techniques may be used.

2.2   How Do I Transfer My Introduced Sample into the Analytical Column?

Usually, the head of the GC analytical column is installed into a GC inlet which facilitates the transfer of the sample onto the stationary phase, see Figure 2.1. The GC inlet is also known as the GC injector, however, the word injector can also be used for the autosampler and therefore we will use 'inlet' in this book to avoid confusion. For some GC set-ups an inlet is not used, and the GC column may be installed directly into the autosampler or valve, for example with some thermal desorption (TD) autosamplers or valves with a gas loop, but this is not always the case. Whichever way the sample is introduced it should:

- Introduce the sample into the column without affecting the separation quality of the column, that is in a narrow sample band in which all of the molecules of an analyte are close together.

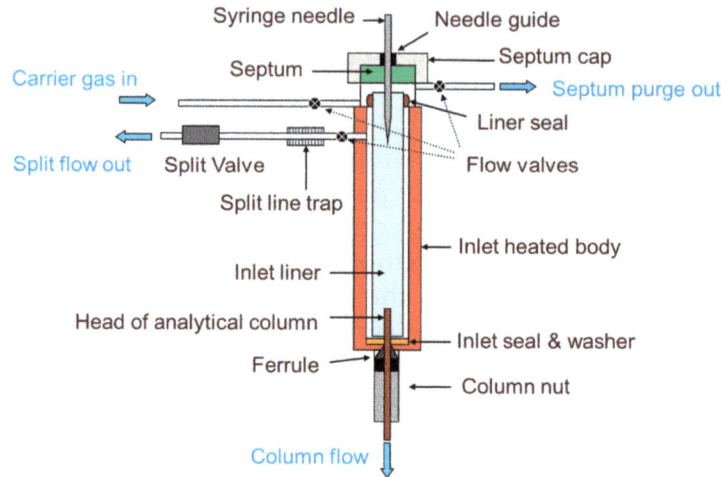

Figure 2.1 A schematic diagram of a typical split/splitless inlet. Courtesy of Anthias Consulting Ltd.

- Without discrimination of the sample components.
- Without thermal degradation, adsorption or rearrangement of the sample components.
- Introduce the sample in quantitative levels and be reproducible for both the trace and major components of the sample.

It is most preferable that any changes in the chromatograph operational conditions should not influence the sampling process. There is currently no ideal sample inlet that meets all the requirements described above for GC and the choice must be based on the requirements for a particular analysis, as well as considering the chemistry of the analytes and matrix.

The two types of GC injection techniques are volatilisation (hot) and cold injections. In volatilisation injections, the sample and solvent are instantly vaporised upon injection as the inlet is maintained at a constant high temperature. In cold injection, the sample and solvent are injected at a relatively cool inlet temperature and then heated to vaporise them. The definition of cold injection is one in which the temperature is below the boiling point of the solvent, whereas a hot injection is above the solvent boiling point and the solvent immediately vaporises.

When the liquid is introduced into the hot inlet containing a liner, the liquid expands in volume. The exact volume is dependent on the liquid introduced and the temperature of the GC inlet, care should be taken to not "overfill" the GC inlet with solvent as this causes a variety of problems.

2.3 How Do I Select Which Inlet Is Suited for My Application?

There are many types of GC inlets such as split/splitless and direct interfaces for vaporising injections; programmable temperature vaporisers (PTVs) and cool on-column inlets for cold injections. Our discussion will focus on the three popular inlets: cool on-column, split/splitless and PTVs. When selecting the best sample injection technique for a particular application, whether on-column, split or splitless, the concentration levels of the components in the sample, their boiling point ranges and thermal stabilities, along with the nature and properties of the solvent all have to be considered.

2.3.1 On-column Inlets

The cool on-column injection, usually of liquid samples, is performed by injecting the sample directly into the analytical column using a narrow syringe needle. The head of the column is positioned towards the top of the inlet and the inlet is held at a temperature of 10–20 °C below the boiling point of the solvent (hence it is known as a 'cold' injection). Once injected, the inlet and the column oven are heated together to first evaporate the solvent and then perform the separation. On-column injection is useful for:

- Thermally labile compounds, as the sample is injected at a cool temperature before being gently heated.
- High molecular weight compounds (up to C100 or more), as they are not subject to mass discrimination when being injected or transferred into the column.
- Trace level analytes, as they are far less likely to suffer from losses resulting in better reproducibility.

However, on-column injection is unsuitable for dirty samples, as the unwanted matrix is also transferred to the column. It is also limited by the column internal diameter (i.d.), solvent-stationary phase polarity incompatibilities and wide sample bands. Therefore, a retention gap (also known as a guard column or pre-column) is frequently used with on-column injections to:

- Protect the analytical column from the matrix.
- Enable on-column injection into narrow-bore columns as the retention gap has a wider i.d. of 0.32 or 0.53 mm.
- Focus low boiling point analytes which are spread across a wide sample band, as retention is stronger on the analytical column compared to the uncoated retention gap.
- Where there is a mismatch between the polarity of the solvent and that of the analytical column causing droplet formation and leading to split peaks, the polarity of the retention gap can be chosen to match the solvent and the analytical column can be selected to match that of the components to be separated.
- To enable large volumes of sample to be injected, see discussion later in this chapter (Section 2.3.3.2).

A programmable temperature vaporiser (PTV) (See Section 2.3.3 in this chapter) can also be used for on-column injections by using an inlet liner insert that holds the head of the analytical column in position for the syringe needle to be inserted.

To summarise, on-column injections:

- Are ideal for high molecular weight, thermally labile and trace labile analytes in clean samples at concentrations up to 50 ppm.
- Suffer problems from dirty samples and solvent compatibility.
- Use of a retention gap can help in several ways.

2.3.2   Split and Splitless Inlets

Split and splitless methods are types of vaporising injections, the sample is in the gas-phase before it is transferred into the column, leaving behind the dirt and unwanted matrix usually in a plug of glass or quartz wool held within the inlet liner or deposited on the liner inner walls. The removable inlet liner can then be replaced before the dirt causes too many problems. The sample is injected into the top half of the liner, the inlet heated and the vaporised sample is transferred into the head of the analytical column, which is positioned close to the bottom of the inlet liner, see Figure 2.1. The selection of either split or splitless mode is determined by the operation of the inlet valves.

2.3.2.1 Split Injection

During split injection mode, the split valve is open and the majority of the sample injected goes out through the split exit, that is to waste and the smaller fraction is directed into the head of the capillary column. As an example, for a split ratio of 100:1, 100 parts goes to waste and 1 part enters the column and is analysed. If the column flow is 1 mL min^{-1} then the split flow will be 100 mL min^{-1}. Split injections usually produce sharp peaks as there is a high flow through the inlet liner (column flow plus split flow = 101 mL min^{-1} in the example) resulting in short sample transfer times to the column. A schematic of a typical hot split injection with a split ratio of 100:1 is shown in Figure 2.2.

To summarise, higher split ratios result in:

- Less sample transferred to the column for analysis and therefore reduced sensitivity.
- Higher flows and therefore faster transfer of the sample into the column producing sharper sample bands resulting in better signal to noise ratios.

Lower split ratios result in:

- More sample on the column for analysis, which is better for low concentration samples but can result in peak overloading of high concentration samples.

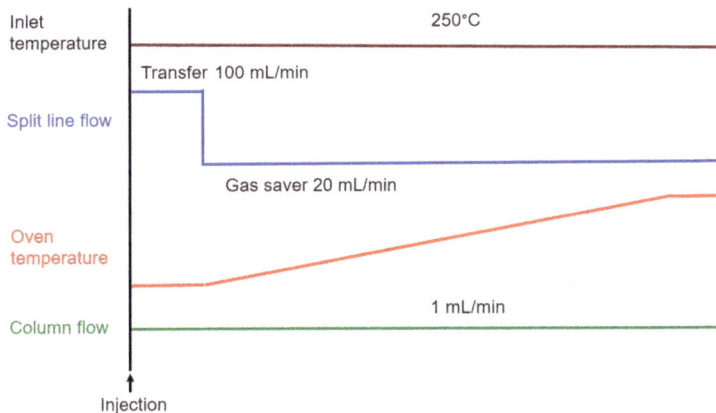

Figure 2.2 A schematic representation of a hot split injection. Courtesy of Anthias Consulting Ltd.

- Slower transfer of the sample to the column which can result in broader sample bands and broader peaks for volatile analytes that do not focus well on the column, resulting in lower signal to noise ratios.

Therefore, it is a trade-off between the sensitivity and peak shape!

In most modern GC instruments the split flow is usually calculated by the software, taking into account the column diameter and length, inlet and outlet carrier gas pressures, column temperature and the carrier gas type to determine the column flow; then determining the required split flow from the split ratio that is entered in the method. Those parameters that need to be entered manually, for example column dimensions, must be correct in order to obtain an accurate calculation of the required split flow to acquire the requested split ratio. If MS is used as the detector, the column outlet pressure will be zero as the MS is under vacuum, this option is usually configured in the software and used in the calculation.

In split injection, mixing of the sample vapour and carrier gas to obtain a homogenous mixture is required before the split occurs at the head of the column. The use of a quartz wool plug, ideally located in the middle of the liner, or fritted or baffled liners is required to achieve this, particularly for high split ratios, to produce good mixing. In contrast, with splitless injections, for concentration of the sample in the column, the mixing of the carrier gas and sample vapour should be suppressed to keep the sample band as narrow as possible. Narrow liners, avoiding wool plugs if possible, are used to give minimal sample dilution in the stationary phase.

To summarise, split injection is ideal to:

- Analyse higher concentration (not trace) analytes greater than 50 ppm, so that the column is not overloaded.
- Produce good peak shapes (particularly at the front of the chromatogram), resulting in narrow peaks which may be better resolved.
- Analyse more active compounds. The fast sample transfer gives less time for interactions, producing less activity effects and improving their repeatability and reproducibility.

2.3.2.2 Splitless Injection

The sample is introduced into the heated inlet liner with the split valve closed. The majority of the evaporated sample flows into the column for a period of time known as the splitless time. After the splitless time or at the purge open time, the split valve is opened and the liner is flushed to remove residual vapours from around the outside of the liner and through to the split valve. This ensures they do not gradually purge into the column throughout the run, which would lead to tailing and ghost peaks.

With splitless injections:

- The entire sample (or the majority) injected into the inlet is transferred to the column using a tapered liner that helps to direct the molecules into the head of column.
- It is used for trace analysis as high concentration samples would overload the analytical column.

Unlike in split injection, transfer of the vaporised sample to the analytical column is relatively slow as the split flow is 0 mL min^{-1} making the flow through the liner equal to the column flow. To ensure that the band broadening effects on the eluent bands from the vaporised sample in the heated inlet are minimal, two re-focusing methods are used: cold trapping and the solvent effect. The column oven temperature is initially set low to enable condensation of the less volatile analytes and they are concentrated close to the head of the column just below the inlet at the end of the temperature gradient, by cold trapping with no migration down the column. To focus more volatile analytes the solvent effect must be used. The initial oven temperature is set 10–20 °C below the solvent boiling point so that the solvent is recondensed close to the head of the column trapping the more volatile analytes in the thin liquid film on the walls of the column. As the oven temperature increases, the solvent evaporates from the inlet end carrying and concentrating the volatile analyte molecules into a sharp band. If there is a polarity mis-match between the solvent and the stationary phase, the solvent will not form a continuous liquid film, but will condense as droplets, such as if you dropped oil onto water. This will result in multiple sample bands of these volatile analytes, producing split or ghost peaks. As with on-column injection, a retention gap of the same polarity as the solvent can be used. A schematic of a typical hot splitless injection is shown in Figure 2.3.

For thermally labile and active compounds, the slow transfer out of the GC inlet in a splitless injection can result in more pronounced thermal degradation or activity. Therefore, a hot splitless injection for thermally labile compounds may not be the best choice. For active compounds, deactivation of the inlet liner and any packing material is required. For unclean samples, quartz wool is less active than glass

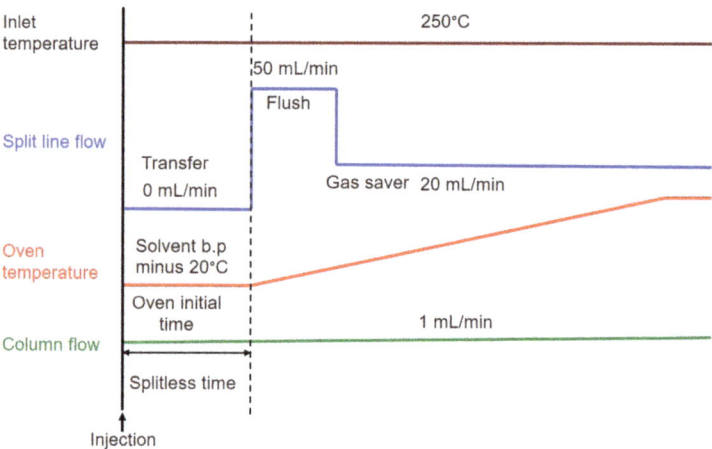

Figure 2.3 A schematic diagram of a hot splitless injection. Courtesy of Anthias Consulting Ltd.

wool and a small plug just above the liner taper, and hence the head of the column is the best position to trap dirt whilst minimising activity. Where possible, it is best to avoid using any wool for clean samples containing trace-level active analytes.

To summarise, splitless injection:

- Is ideal for analysing low concentration and trace analytes of less than 50 ppm, as the entire introduced sample is transferred onto the column.
- Owing to the slow transfer process to the column, cold trapping and the solvent effect may need to be relied upon to produce good peak shapes.
- To analyse more active, trace compounds, activity must be reduced in the sample flow path otherwise losses will occur, deactivation is important.
- The degradation of thermally labile analytes may be more pronounced.

2.3.3 Programmable Temperature Vaporiser Inlets

The PTV inlet closely resembles the split/splitless inlet, but usually uses narrower inlet liners and has precise temperature control to linearly or ballistically increase the inlet temperature and also to rapidly cool it, all operations are controlled through the software or keypad. PTVs can be used for regular hot split/splitless injections, although be wary of lower PTV liner volumes when compared to those of a split/splitless liner and, for cold on-column injections when using a special on-column liner. However, they also have many other modes of operation including cold split and splitless methods, large volume injections and some enable one-shot thermal desorption and pyrolysis analyses to be performed, depending on the liner size, maximum heating rate and maximum temperature. Here we will consider PTVs for cold and large volume (LVI) injections.

2.3.3.1   Cold Split and Splitless Injections

Split/splitless inlets can only be set at one temperature which is selected to vaporise all of the compounds of interest, but this high temperature can cause mass discrimination in the syringe or thermally degrade unstable analytes. A PTV enables a cold injection to be performed (as in on-column injections) but the analytes are vaporised before transfer to the column, enabling dirty samples to be injected and the matrix to be retained in the liner. This technique is useful for samples that are not clean enough for on-column injection but contain:

- High molecular weight analytes
- Thermally labile analytes.

The sample is introduced as a liquid into a cold liner and coats either the inner walls or any packing material present. The liner temperature is subsequently raised to a normal operational temperature at a rate that is suitable for the analytes in the sample.

In cases in which the boiling point of the analyte is sufficiently higher than that of the solvent, large sample volumes can also be injected with PTV inlets in LVI mode. A schematic diagram of a typical cold splitless injection is shown in Figure 2.4.

2.3.3.2   Large Volume Injections

Rather than concentrating the sample manually, for example by evaporating the solvent under a nitrogen flow in which the conditions may be less controlled and volatiles could be lost, PTV inlets enable sample concentration to occur in the inlet under more controlled conditions. Split and splitless injections are limited, usually to around 3 µL, by the liner volume and the solvent band length when recondensed on the column, but this is dependent on the size of the inlet and the column. In solvent vent LVI, a much larger volume of solvent, from 10 µL up

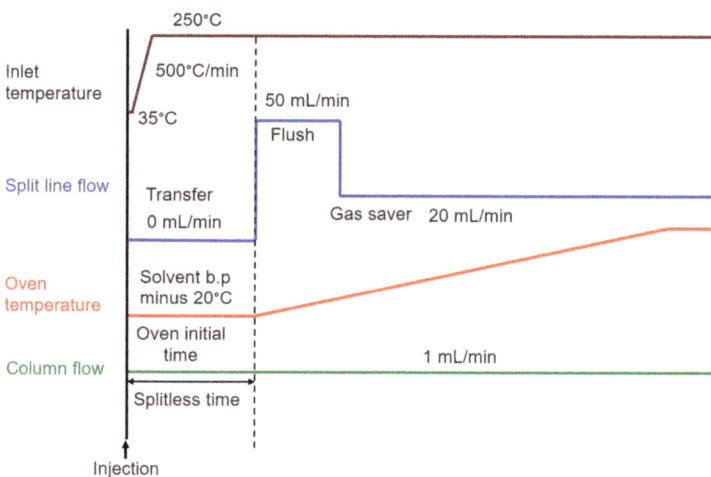

Figure 2.4 Schematic diagram of a cold splitless injection. Courtesy of Anthias Consulting Ltd.

to 1000 μL, is injected into a cold inlet and the evaporated solvent is vented *via* the split exit while components of the sample are trapped and concentrated within the inlet liner. Then, the split line valve is closed and the temperature is increased to transfer the concentrated sample into the column in splitless mode, in which the analytes are reconcentrated by cold trapping and using the solvent effect.

There are three types of solvent vent methods, and their use is dependent on the type of PTV, the size of the inlet liner, the type of autosampler and the volatility of the analytes in comparison to the solvent.

- Rapid or at-once injection uses a packed liner which absorbs all of the sample injected (around 10–150 μL), in that solvent and at that specific temperature and pressure. If any changes are made to the parameters, that volume may exceed the capacity. The challenge is to find a packing material that does not cause analyte loss but absorbs the full sample volume required for the method. This technique is usually used with mid-sized PTVs, for which the liner can hold a good amount of the packing material, but it can all be rapidly and fully heated for fast transfer to the column. The biggest advantage is that when the solvent is evaporated a cold spot is formed in the middle of the packing, this reduces the loss of volatile components. The temperature of the cold spot is dependent on the solvent type. With an initial inlet temperature of 30 °C, the evaporation of pentane will reduce the packing temperature to below −15 °C,[1] whereas methanol would only reduce it to around 10 °C, see Figure 2.5. For example, if hexane is the

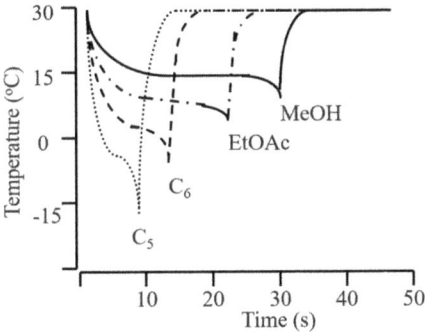

Figure 2.5 Temperature decrease within PTV inlet for LVI of different solvents. From ref. 1 courtesy of GLSciences.

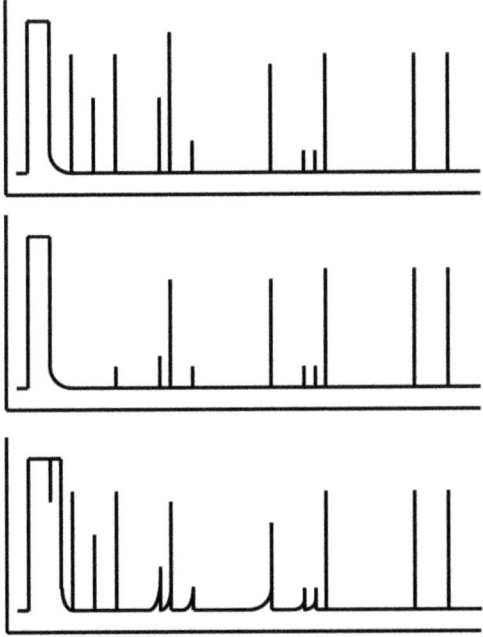

Figure 2.6 LVI optimisation: ideal chromatogram (top), when the solvent elimination time is too long (middle), and when the elimination time is too short (bottom). From ref. 1 courtesy of GLSciences.

solvent, excellent recoveries can be obtained from C8 upwards with the cold spot dropping to −5 °C. If the liner is evaporated to dryness the solvent and the cold spot disappear leading to volatile losses, therefore the split exit should always be closed with 1–3 µL of solvent remaining in the liner. Too much solvent being transferred leads to a long solvent band on the column resulting in poor peak shapes for mid-volatility analytes, see Figure 2.6.

- The speed controlled LVI technique is useful when the volume to be injected exceeds the liner capacity. The sample is introduced at a constant and controlled rate with the split valve open and the solvent evaporation rate approximately equal to the injection speed. Ideally, the injection speed should be slightly faster than the evaporation rate, so that a small volume of solvent accumulates within the inlet liner to reduce losses of more volatile compounds, as described earlier. However, the volume should not exceed the liner maximum liquid volume, a small amount of packing can help! Baffled liners are frequently used for these injection types. After injection, excess solvent is evaporated (but not to dryness), the split exit closed and a splitless transfer to the column is performed. The maximum sample volume is unlimited, up to 1000 µL of less volatile analytes has been injected, however the solvent evaporation takes a long time. As the formation of a cold spot is limited, very volatile analytes close to the boiling point of the solvent are lost, for example in hexane only analytes above C12 have excellent recoveries. Depending on the manufacturer, calculators exist to help determine the starting point for the injection speed. Too fast and the liner capacity is exceeded, too slow and more volatile analytes are lost. An autosampler must be available in which the precise injection speed can be controlled.
- The multiple injection LVI technique makes multiple (as many as required) fast injections of a fixed volume (usually around 4 µL, but this is liner dependent), with the split valve open. Solvent evaporation and sample concentration takes place between each injection. After the final aliquot is injected and any excess solvent is evaporated, the split exit closed and a splitless transfer is performed. As in the previous described techniques, evaporation to dryness results in a loss of volatiles, but this is more enhanced as multiple evaporations are taking place therefore, only C14 and above give excellent recoveries in hexane. The possibility of flooding the liner with too much residual solvent after evaporation is also higher, again a small amount of packing material helps. However, the ability to perform this technique is not limited by the liner capacity or the precise injection speed, but the ability of the autosampler to perform multiple injections.

Large sample volumes can also be injected with the on-column technique. The retention gap must be sufficiently long to hold the volume injected, the rule of thumb is 1 µL = 20 cm + a safety margin. For example, for a 100 µL injection this would be 20 m + 5 m = 25 m retention gap! This results in slow solvent evaporation and therefore long run times; there is a much larger amount of matrix being injected (in this example 100*x* therefore the sample has to be clean), plus all of the solvent passes through the detector and can reduce the sensitivity. A method that can be used to prevent this is to use an early solvent vapour exit. A valve is placed between the retention gap and the analytical column, the evaporated solvent is directed out of the exit and not through the analytical column. The timing is then critical to switch the valve and direct the analytes into the analytical column, which must be adjusted when the retention gap is trimmed.

To summarise, LVI:

- Enables better detection limits to be reached or a manual concentration step to be avoided.
- Is a cold injection technique and therefore useful for all analyte types.
- Solvent vent methods trap the matrix in the inlet liner, but on-column LVI is less resilient and can only be used with clean samples.
- There are more parameters and more steps to optimise, however put simply, it is a splitless injection with a pre-step of solvent elimination.
- Calculators can help to optimise the parameters.
- Loss of volatiles at the front of the chromatogram means the liner has evaporated to dryness and solvent evaporation should be reduced.
- Poor semi-volatile (C14 to C16 region) peak shapes mean that too much solvent has been transferred onto the column and therefore greater solvent evaporation needs to occur.

2.3.3.3   In-liner Sample Preparation

Most PTV inlets not only enable the inlet to be cooled between sample analyses, but also enable high control of the split valve, split flow, inlet temperature ramp rate and the final temperatures, which usually extend above the maximum of most split/splitless inlets. Inlet programs can be created with multiple temperature ramps and split flows. This enables sample introduction to be more creative, for example in-liner derivatisation to be performed.

The sample may already be in the liner (see liner exchangers in Section 2.3.3.4) or may be co-injected with the derivatisation reagent into a cool inlet under stopped-flow conditions (no or minimal liner flow), with the column head pressure turned off. In some methods it may be necessary to evaporate excess solvent, for example methanol, to prevent the liner volume being exceeded, this is performed by following the steps used for LVI. After solvent evaporation, the concentrated sample and reagent(s) are then rapidly heated to a set temperature for the reaction to occur. The final step is to turn on the gas flow to flush the products into the GC column usually under splitless conditions.

To summarise:

1. Inject sample and reagents into the inlet liner with the inlet pressure turned off or minimal.
2. If necessary, evaporate excess solvent with the split valve open, using similar parameters to LVI.
3. Close the split exit and rapidly heat the inlet to perform the reaction.
4. Turn on or increase the inlet pressure and transfer the products onto the column in splitless or split mode.

2.3.3.4   Liner Exchangers

Another advantage of PTV inlets, when fitted onto a GC with specific autosamplers, is that an additional liner exchanger can be installed. This can enable the inlet liner to be automatically changed as frequently as every sample. The inlet (and GC oven) are cooled and the inlet pressure is reduced or turned-off before the old liner is removed. The new liner is installed, pressurised and flushed before the next sample (or a blank) is injected. This can be useful for the analysis of:

- Very dirty samples that require the inlet liner to be changed very frequently, for example every 10 injections.
- Samples prepared in the inlet, for example with in-liner derivatisation for which a clean liner may be required for each sample.
- Samples treated using thermal desorption, thermal extraction or pyrolysis, with each liner containing a different sample and stored in a liner rack on the autosampler.

2.4   Comparison of Injection Techniques

Table 2.1 below provides an overview of the different injection techniques and compares the temperatures, maximum volume, sample type and types of analytes that are suitable.

Table 2.1 A comparison of injection techniques.

Technique	Temperature	Maximum volume	Samples	Analytes
On-column	Cold	0.1 µL+ (retention gap dependent)	Clean only	Trace, labile and high MW
Split	Hot or cold	0.1–3 µL (liner dependent)	Clean to dirty	High concentration
Splitless	Hot or cold	0.1–3 µL (liner dependent)	Clean to dirty	Trace
Large volume	Cold	3 µL+ (liner dependent)	Clean to dirty	Very low

2.4.1 How Do I Select and Optimise My Inlet Method Parameters and Consumables?

Many people think that there are not many parameters to consider or optimise in GC sample introduction and many methods can get away with standard values being used. However, to obtain the most robust method, particularly for more difficult analytes and samples, each step of the process should be considered. Internal standards that are representative of the volatilities and activities of the analytes are very useful for improving precision. The inlet is one part of the GC in which multiple operations take place, even in a simple hot split injection these steps are:

1. Get all of the sample out of the syringe and into the inlet liner reproducibly and without discrimination.
2. Vaporise all of the analyte molecules but minimise the heavier matrix evaporation, this is selective discrimination.
3. Transfer the vaporised sample into the analytical column reproducibly and without unwanted discrimination.

2.4.2 Injection Volume

The volume that can be injected into the inlet and transferred into the column depends on the inlet type, liner volume, column dimensions and injection mode.

Liquid volumes for split and splitless injections are usually between 0.1–3 µL. For hot injections, if the volume is too large, the vapour volume may exceed the capacity of the inlet liner, causing contamination of the system around the outside of the liner, down the split, septum purge and carrier gas-in lines. This will result in non-blank blanks, ghost peaks, poor peak shapes and a reduction in sensitivity and fixing this will take hours of maintenance time. The vapour volume depends on the solvent type (boiling point, density and molecular weight), volume to be injected, inlet temperature and pressure on injection. Calculators[2] can be used to determine this, the values are entered (if a solvent mixture is used, enter the most polar) and the result is compared to the liner volume. If this is not known, the liner volume can be calculated using eqn (2.1) shown below:

$$\text{Liner volume (mL)} = \pi \times \text{radius}^2 \text{ (cm)} \times \text{length (cm)}$$

or

$$\text{Liner volume (\mu L)} = \pi \times \text{radius}^2 \text{ (mm)} \times \text{length (mm)} \tag{2.1}$$

The vapour volume should not exceed 75% of the liner volume to allow for errors. If it does, the injection volume can be reduced or a pressure pulse applied during injection to reduce the vapour volume. Again, the calculator can be used to determine the pressure required, as too high a pressure can result in problems during injection, particular if the septum does not seal around the syringe needle well, and also increases the boiling point of the analytes. More polar solvents result in larger vapour volumes. For example, at 250 °C and 7 psi, a 1 μL injection of:

- Hexane = 222 μL
- Methanol = 719 μL
- Water = 1610 μL

A typical 4 mm i.d. splitless liner has a capacity of 900 μL. It should be noted that water has a very large vapour volume, which is one of the reasons it is difficult to use as a sample solvent for GC. This should be remembered when injecting solvents that could be contaminated with water, as even a trace amount of water could increase the vapour volume to exceed the liner capacity and result in reduced sensitivity for that injection and carryover into the next injection. Therefore, consider drying the sample solvent first, for example with sodium sulphate, but this step should be validated to prevent potential loss of analytes.

For cold injections and LVI, the maximum sample volume is limited by the liner and packing surface area and the affinity of the solvent to it. It is higher when the polarities of the solvent and the packing or liner walls match so that optimal wetting is obtained. For example, when injecting hexane into a cold liner, it runs down the sides and out of the bottom, whereas methanol interacts and coats the inner walls much more effectively. Samples running into the bottom of the inlet can result in ghost peaks and poor recoveries owing to analyte molecules transferring to the column later than the molecules held inside the liner or liquid sample being lost through the split exit. Baffled liners and those with packing can increase the capacity of the liner.

Not only must the liner be considered for the maximum volume injected, but the column too. For any injection in which the oven temperature is below the solvent boiling point, but particularly in splitless transfers in which the solvent effect is used to obtain good peak shapes for more volatile analytes by condensing the solvent on the walls at the head of the column, this is important. If too much solvent is transferred onto the column wide sample bands can result, which produce poor peak shapes for mid-volatility analytes, see Figure 2.7. Involatile analytes, with boiling points greater that 150 °C above the

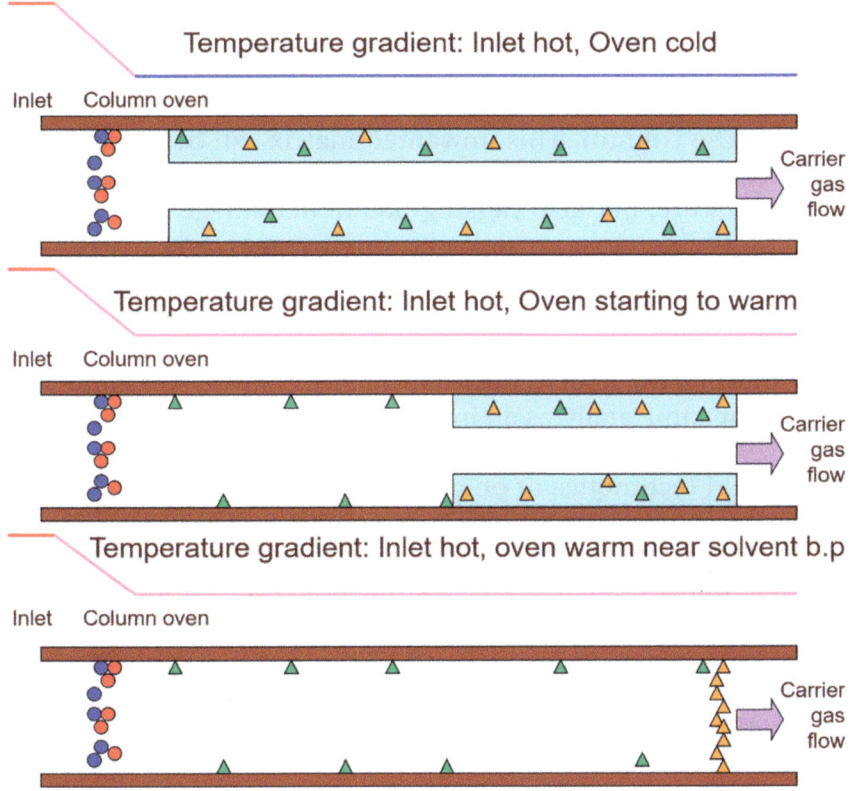

Figure 2.7 The solvent effect and cold trapping: cold trapping of the involatiles (red and blue) and recondensed solvent band (top); evaporation of the solvent and focusing of the volatiles (yellow) (middle); focused involatiles (red and blue) and volatiles (yellow), semi-volatiles (green) as a broad band (bottom). Courtesy of Anthias Consulting.

solvent boiling point are cold trapped. Volatile analytes, with boiling points less than 50 °C above the solvent boiling point remain dissolved in the solvent and move with it when it evaporates to concentrate in a narrow band. The semi-volatile analyte molecules, with boiling points 50–150 °C above the solvent boiling point are left spread-out along the walls of the column from which the solvent has condensed, as a wide band and do not focus, resulting in broad and distorted peaks. Transferring 3 μL or less of solvent to the column minimises this effect. For larger volumes, a different focusing mechanism, for example a retention gap must be used.

2.4.3 Temperatures

The inlet temperature for hot split and splitless injections, is dictated by the volatility of the analytes of interest. The inlet should be hot enough to obtain total transfer of the least volatile analyte of interest; while keeping it as low as possible, to minimise the likelihood of any thermal degradation and minimise the transfer of any unwanted matrix into the column. This unwanted matrix can be more difficult and time consuming to remove, increasing the run times and the final oven temperature, but also causing potential damage to the column and soiling the detector. An inlet liner is cheaper, and is easier and faster to replace. Start with a higher temperature (320/300/250 °C depending on the analyte volatility and the maximum isothermal temperature of the column installed into the inlet) and gradually reduce it, while plotting the area of the least volatile analyte. No temperature is too low, for some analytes an inlet temperature of 40 °C is adequate to vaporise them, hence the need to carefully optimise!

For the cold techniques of on-column, cold split and cold splitless injection the initial inlet temperature should be 10–20 °C below the solvent boiling point to ensure that the solvent stays in the liquid phase upon injection. For a large volume injection it should be 10–30 °C below the solvent boiling point. The temperature is then increased depending on the analytes and the PTV used. For thermally labile analytes 1–4 °C s^{-1} should be used, for thermally stable analytes the maximum heating rate can be used, if in doubt, use a moderate rate of around 8 °C s^{-1}. The final temperature is determined using the same method as for hot injections.

The initial oven temperature should be selected and held while all analyte molecules are transferred onto the column, otherwise band broadening will occur. For a splitless injection the oven initial hold time should be equal to the splitless time. For split injections with a very fast transfer, the solvent effect is not required and therefore the oven starting temperature depends on the volatility of the first eluting analyte. For cold on-column, hot and cold splitless injections the initial oven temperature must be 10–20 °C below the boiling point of the solvent for the solvent effect to occur and it is useful to optimise this, particularly for LVI as it effects the length of the solvent band.

2.4.4 Sample Transfer Time to the Column

For splitless injections the splitless time should be long enough for complete transfer of the target analytes onto the column. Opening the split valve too early will result in sample loss through the split exit which will produce reduced sensitivity and lower reproducibility. However, it should not be too long either, otherwise the solvent and analyte molecules will start to migrate down the column resulting in broader and tailing peaks. The time taken to transfer the sample into the column mostly depends on the volume and flow through the liner and should not take longer than two flushes of the liner volume. The inlet liner volume may be stated on purchase or can be calculated using eqn (2.1). Ignoring any tapers it will usually be between 0.2 and 1.2 mL depending on the inlet type and manufacturer. The time taken for two flushes of the liner volume can then be calculated using eqn (2.2):

$$\text{Splitless time (min)} = \left(\frac{\text{liner volume (mL)}}{\text{liner flow (mL/min)}} \right) \times 2 \qquad (2.2)$$

In which the liner flow is usually equal to the column flow (as split flow = 0 mL min^{-1}) unless a pressure pulse is used. From this starting value (which can be rounded up, *e.g.* 1.8 to 2 min) the maximum time taken to transfer the sample can be calculated. This can then be optimised to shorter times if necessary. Transfer time (and so oven initial hold time) can also be calculated in the same way for split injections, however the liner flow will equal the column plus split flows.

2.4.5 Split/Vent Flows

In a split injection the split flow is dependent on the split ratio required and the column flow, to transfer a suitable portion of the sample onto the column for analysis depending on the concentration. It is always best to start with a higher split ratio than is believed to be needed, to ensure a high concentration is not transferred onto the column, which could cause contamination. The split flow can then be reduced in order to optimise for sensitivity.

In a splitless injection the split flow is used to clean the liner. With higher split flows, better evaporation will occur for the less volatile analytes, also a smaller portion will transfer onto the column with the majority going to waste. However, a higher split flow also results in more gas usage. If the GC is equipped with a gas saver mode, enabling the split flow to be reduced after a fixed time, then a high split flow (*e.g.* 100 mL min^{-1}) for a short period of time (*e.g.* 2 min) will effectively clean the liner, before using a gas saver mode of 15–20 mL min^{-1} flow. If the instrument does not have a gas saver mode, a split flow of 30 mL min^{-1} is a good compromise to clean the liner, but to also conserve gas.

In a large volume injection the vent flow (split flow when evaporating the solvent) is dependent on the volume injected and the type of solvent. Less polar solvents, for example hexane, evaporate very quickly, whereas more polar solvents, for example methanol, are much slower and therefore a higher vent flow is preferable. Ideally, solvent evaporation will take between 15–60 s with vent flows of 50–250 mL min^{-1}. Shorter periods, less than 15 s, result in the solvent evaporation being too fast, and reproducibility and accuracy are then difficult to achieve. Long periods, more than 60 s, waste analysis time, however for speed-controlled injections or when very large volumes are injected or very polar solvents used, solvent elimination can take a long time, for example 10 µL of water in a packed liner can take 10 min! The rule of thumb is that the evaporation rate doubles when the vent flow is doubled.

2.4.6   Liner Size and Style

The liner size, length and outer diameter (o.d.), are dependent on the manufacturer and the type of inlet. For example, liners for split/splitless inlets tend to be larger in order to accommodate the solvent vapour volume, whereas for PTVs they tend to be shorter and have a narrower o.d.

The i.d. is selected depending on the injection technique used and the vapour volume. For a hot vaporising injection of a polar solvent the liner volume needs to be as large as possible to accommodate the large vapour volume. For a hot vaporising injection of a non-polar solvent a narrower liner can be used, which more closely matches the vapour volume, resulting in a faster transfer of the sample to the column with a shorter flush and therefore a shorter splitless time. For a solventless injection, for example solid-phase microextraction (SPME), a very narrow liner is used to facilitate a very fast transfer to the column. In the case of SPME a 0.75 mm i.d. liner is optimal and using wider i.d. liners would result in poor chromatography. A headspace injection is a vapour injection, not solventless, and a liner volume must be selected that matches the volume injected at the injection speed.

Again, the style is selected to match the technique. Splitless transfer is most efficient when the vapour is directed into the head of the column, for example with a taper. A taper towards the top of the inlet liner helps to prevent flashback towards the septum, however it also reduces the liner volume and if the syringe needle style, the inlet needle penetration and the vapour volume are correct a taper is not required. Split injections require rapid mixing of the vapour with the carrier gas which is achieved using baffles, a frit or wool located in the centre of the liner. The matrix is best trapped with a plug of wool at the base of the liner. If the liquid injected is viscous, the tip of the needle can be wiped with the wool positioned towards the top of the inlet liner.

2.4.7   Liner Packing Material

The packing must be carefully selected for the greatest recovery of analytes with no irreversible adsorption, thermal or catalytic breakdown. It should also give good blanks, not be damaged by the solvent and be stable at the final inlet temperature. This is very important for the packed liners used in rapid LVI.

Activity is the biggest problem for active compounds, causing tailing peaks, irreversible ab/adsorption or breakdown, all of which result in reduced sensitivity and inconsistent results for those analytes. Ensuring both the liner and the packing material will not interact is important, particularly for trace level analytes. The most common packing material is glass wool, however, even if this is deactivated the breakage of any fibres results in an active surface. Therefore, if packing your own liners, it can be preferable to buy glass wool that is not deactivated, then once the whole liner is packed, including any packing, it can then be deactivated. Alternatively, quartz wool can be used which is naturally less active. Glass beads and quartz tubes have also been used, particularly for LVI. Glass frits and baffles increase the surface area of the liner and are more easily deactivated. Polymer packings are popular for LVI as they are more adsorbent, however they can be more active for certain functional groups, generally have lower temperature limits and certain solvents can damage them.

When selecting a packing material, activity can be checked by analysing standards using an empty, deactivated liner and comparing recoveries to the packed liner.

- First, a conditioning standard should be injected, and it is useful to do this whenever a new liner is installed. This could be a high concentration standard, preferably in the matrix, to mop up any active sites (there are always some, even with a deactivated liner).
- Next, one or two solvent blanks should be injected to flush and clean the liner.
- A low concentration standard should then be analysed to check for activity.

2.4.8   Deactivation

Every surface in the sample flow path can potentially cause problems for active compounds, whether it is glass or metal. The inlet is the main source of activity, particularly after many injections, when the matrix builds-up in the liner, but this can only be fixed through maintenance. Using a fully deactivated system for these types of analytes is important. Deactivation applies an inert, integral layer to the surface.

- For glass, the active silanol (-Si-OH) group is derivatised most commonly by silanisation with, hexamethyldisiloxane for example. However, this is reversible through damage or temperatures above 400 °C.
- Siltek™ is a passivation process which is more stable at temperatures up to 450 °C.[3] It is mostly used for inlet liners, giving a shiny brown appearance, but can also be used for press-fit connectors and high nickel alloys of steel.
- Silcosteel™ is a general purpose passivation layer for steel or stainless steel that is used throughout the system from the thermal desorption tubes to internal plumbing tubes, metal liners and the entire inlet body. The maximum temperature is 600 °C.

For active compounds the following parts of the inlet should be considered:

- Deactivated liner: whether the liner is made of glass or metal it should be deactivated to reduce the surface activity.
- Deactivated liner packing material.
- Deactivated inlet seal at the base of the liner, used in Agilent gas chromatographs. These can be gold or silcosteel™.
- Deactivated inlet body, particularly useful when analysing highly reactive sulphur compounds.
- Deactivated transfer lines from autosamplers.

2.4.9 Septum

The septum enables introduction of the sample into the inlet liner without pressure loss, by sealing around the syringe needle upon injection, keeping the sample inside and the air and contaminants out of the inlet. There are many variations of septa, from general purpose silicone rubber to butyl rubber, Viton, PTFE-faced and duckbill long-life seals.

When selecting the septa type for the application there are a number of variables to consider, including the inlet temperature, type of solvent and number of injections to perform. Most septa work with most solvents up to a maximum of 350 °C for 200–400 injections. High temperature septa are suitable up to 400 °C, however they are made from a harder material and will start to leak with fewer injections. Some solvents degrade the material of some septa (it transfers into the septum from the outside of the syringe needle) and can reduce the lifespan. For example, the duckbill long-life seals are good for up to 2000 injections but can be damaged by dichloromethane.

For trickier applications, for example if performing in-liner derivatisation, different septa materials should be tested to find the most suitable with minimal degradation. Tetramethylammonium hydroxide (TMAH) derivatization is very harsh, the most resistant septum is the Thermolite septum, but this still must be replaced every 10 injections!

When septa are damaged they bleed. High concentration samples can also be gradually released from the contaminated septa causing ghost peaks and carryover. The septum purge is a continuous flow of carrier gas (1–3 mL min^{-1}) over the bottom surface of the septa and out to waste, to reduce or eliminate these problems. More recent gas chromatographs can now programme this flow to aid with trickier applications.

2.5 How Do I Introduce a Gas into the Gas Chromatograph?

Gas can be introduced into the GC using a gas tight syringe or *via* a gas switching valve. The gas switching valve is a more reliable and robust method than a gas tight syringe but for simple qualitative analysis a gas tight syringe is adequate. It is important, for obvious reasons, to check the syringe is gas tight. Prior to injection it is also important to ensure the syringe is clean and purged of any contamination. This is most readily achieved by flushing the syringe repeatedly with a dry high purity gas such as nitrogen, never use solvent! The injection of samples can then be made into the GC inlet through the septum. Bear in mind that samples are already in the gaseous state so the evaporation step in the inlet is not required and the volume change upon injection is now a function of the pressure of the carrier gas and the temperature of the inlet. Meaning that there is no expansion of the liquid into a gas. This allows higher volumes to be injected directly into the inlet compared to a liquid injection, and volumes up to 500 µL are not uncommon. However, the rate of injection should not result in the capacity of the liner being exceeded, particularly if a splitless injection is used, which has a slow liner flow. As previously described, the calculation should be used to determine the maximum injection rate.

When gases are introduced into the GC *via* a gas switching valve, which incorporates a sample loop, the exact configuration of the switching valve can vary depending on the application, but the basic configuration is similar to that shown in Figure 2.8. In position 1, the valve is switched so that the sample loop is filled with the sample. In position 2, the valve is switched so that the content of the filled sample loop is swept into the GC column by the carrier gas flow.

The gas sampling valve is usually heated, along with the entire sample flow path after this point, for accurate and reproducible injection volumes and to prevent condensation of the sample. Heated valve ovens are used, or the valve can be positioned inside the GC oven. The valves themselves should have a high repeatability, with low dead volumes and no sample carryover. Rotary valves rotate in one direction to load and the opposite direction to inject the sample, they are robust at high pressures and temperatures and are rotated by hand, air-pressure or electric actuators. Slider valves have lower volumes to give narrower initial peak widths and faster switching times, however they are not common as they are not as robust at high temperatures and pressures.

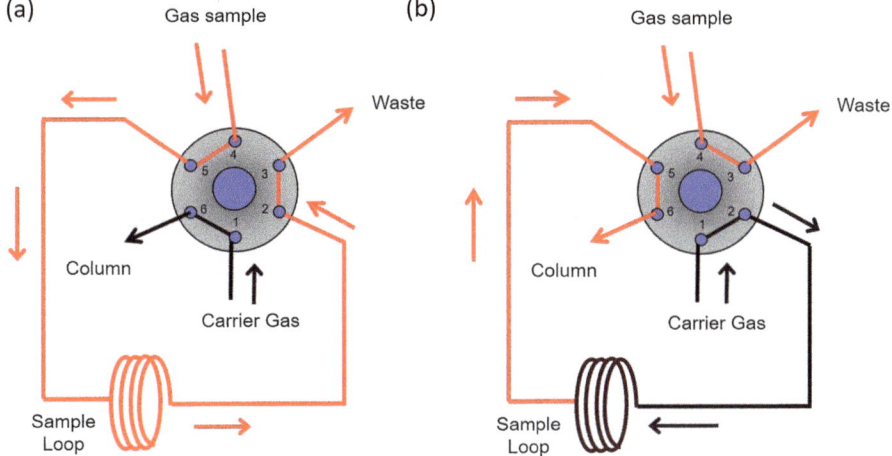

Figure 2.8 A six-port gas valve. Position 1 (a) is filling the sample loop then Position 2 (b) is "injection" into the column.

References

1. ATAS/GLSciences, LVI training manual. Available at https://www.glsciences.eu/optic/LVI-training-manual.pdf, accessed June 2019.
2. Solvent Expansion Calculator. Available at http://www.restek.com/images/calcs/calc_backflash.htm, accessed June 2019.
3. Frequently Asked Questions: Surface Treatments. Available at http://www.restek.com/Pages/faq_surf, accessed June 2019.

3 Chromatographic Separation

3.1 How Do I Assess the Quality of My Chromatographic Separation?

3.1.1 Chromatographic Systems

The systems are comprised of a carrier gas supply, a sampling system, a sample introduction system or inlet, the separating module or column which is placed in a column oven to control the separation temperature, a detection system and lastly a recording system, which today is a computer data acquisition and processing component (Figure 3.1). For gas chromatography, the gas chromatograph (GC) provides conditions required by the column for achieving separation of analyte components without reducing the performance of the column.

The gas required for a GC-MS is only a carrier gas, but if a collision cell is used for GC-MS/MS for example, then a collision gas and possibly a make-up gas is also required. Some GC detectors require makeup gas and/or combustion gases, depending on the type of detector. The purity of the gases used is very important. For GC, a minimum grade of 5.0 (99.999%) is required, whereas for GC-MS the minimum grade is 5.5 (99.9995%). It is necessary to have gas purification scrubbers to remove traces of oxygen, moisture and hydrocarbons just before they enter the GC. In modern instruments, flow controllers in the hardware control the gas pressure and/or flow.

Gas Chromatography-Mass Spectrometry: How Do I Get the Best Results?
By Diane Turner, Mathias Schäfer, Steven Lancaster, Imran Janmohamed, Anthony Gachanja and Jason Creasey
© Diane C. Turner, Mathias Schäfer, Steven Lancaster, Imran Janmohamed, Anthony Gachanja and Jason Creasey 2020
Published by the Royal Society of Chemistry, www.rsc.org

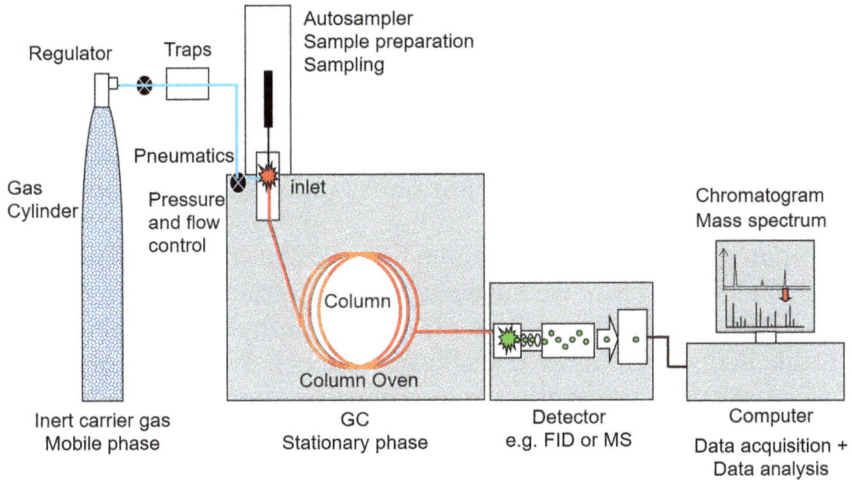

Figure 3.1 Components of a GC with mass spectrometer detector (GC-MS). Courtesy of Anthias Consulting Ltd.

The manufacturers of gas chromatographs include Thermo, Perkin Elmer, Shimadzu, Agilent and others, but what is the difference between gas chromatographs from different manufacturers and how do I choose which GC I need to achieve separations in my laboratory?

3.1.2 Separation Techniques

According to the International Union of Pure and Applied Chemistry (IUPAC), chromatography is defined as a separation process that is achieved by distributing the components of a mixture between two phases, a stationary and a mobile phase. Gas chromatography involves a sample being separated into its component parts by equilibration between the carrier gas (usually helium for GC-MS applications, however nitrogen, hydrogen and argon/methane mixtures are used for many GC applications) and a stationary phase. A component which has a large affinity for the stationary phase will take longer to travel through the column than a component which has a low affinity for the stationary phase. As a result of this, sample components will become separated from each other as they travel along the column. This process is called elution. The processes giving rise to retention are statistical in nature and the rate at which an analyte moves through the column is determined by the average time it spends in the mobile phase. In the best performing systems, the peaks are Gaussian because of the statistical nature of the separation processes.

Therefore, solutes are eluted from the system as local concentrations in the carrier gas in the order of their increasing distribution coefficients with respect to the stationary phase; and *ipso facto*, separation is achieved.

Chromatographic methods can be broadly classified in terms of the mobile phase, therefore we have gas chromatography, liquid chromatography, and supercritical fluid chromatography as well as the kind of equilibria involved in the transfer of solutes between the phases for example partition, adsorption, size exclusion, ion exclusion and ion exchange. This chapter will concentrate on the separations that can be achieved using gas chromatography.

In gas chromatography, the mobile phase is always a gas, note that it is not a phase in the true thermodynamic sense, in that there should be no thermodynamic interactions between the carrier gas and the analyte, unlike the mobile phase in liquid chromatography which solvates the analyte. However the stationary phase can be a solid, a liquid adsorbed onto a solid or an organic species bonded to a solid surface. In gas–solid chromatography (GSC) the solid sorbent is packed into a column, similar to an HPLC column, or affixed to the inner walls of a capillary column as a porous layer open tube (PLOT) column. In GSC, separation is based upon the non-linear adsorption of gaseous substances onto solid surfaces and therefore these columns are used for small gaseous species and volatile hydrocarbons using GC, but not GC-MS instrumentation.

The most common type of gas chromatography, used to separate volatile organic compounds (VOCs), is gas–liquid chromatography (GLC). In GLC, separation is achieved through partitioning of the analytes between the mobile phase and the stationary phase, which is a viscous liquid coated onto the inside of a very narrow tube. This configuration is known as a wall coated open tube (WCOT). The tube itself is a long narrow length of deactivated fused silica, coated on the outside with polyimide to protect the silica and to make it flexible, with internal diameters commonly ranging from 0.1–0.53 mm and lengths that are usually between 10–150 m. The most common dimensions are a 0.25 mm internal diameter and a length of 30 m, enabling fast separations with a good efficiency, peak shapes and sensitivity. GC-MS instruments use WCOT, commonly known as capillary columns and therefore this chapter will focus on these columns, which separate analytes using partitioning.

3.2 How Do I Select the Right Stationary Phase and Column Chemistry?

The stationary phase itself is usually polydimethylsiloxane (PDMS) (Figure 3.2a), which is non-polar. Separation is achieved through dispersion interactions and is based on the volatility of the analytes. The GC oven temperature is started low (usually below 100 °C) and then ramped up to higher temperatures to separate the less volatile analytes.

The dimethyl groups can be replaced with different percentages of different functional groups, to change the interactions. For example, a 5% diphenyl and 95% dimethylpolysiloxane is still deemed a non-polar column but a 50% diphenyl and 50% dimethyl polysiloxane (Figure 3.2b) is a mid-polar column in which separation is based primarily on volatility but also on the π–π interactions between the π bonds of the analyte and those in the phenyl groups of the stationary phase. Other polysiloxane-based stationary phase functional groups include biscyanopropyl, trifluoropropylmethyl and cyanopropylmethyl-phenylmethyl, with interactions that are dependent on the characteristics of the functional groups present in the analytes and those on the stationary phase. These interactions include dipole–dipole and dipole-induced dipole interactions, as well as dispersion and π–π interactions. Another common class of capillary column stationary phase is the unsubstituted polyethyleneglycol (PEG) (Figure 3.2c) also known as a 'wax' column, which enables the additional interaction *via* hydrogen bonding of the –OH group with relevant functional groups in the analyte molecules. Variations of the PEG column are: FFAP, nitroterephthalic-acid-modified for volatile fatty acids and

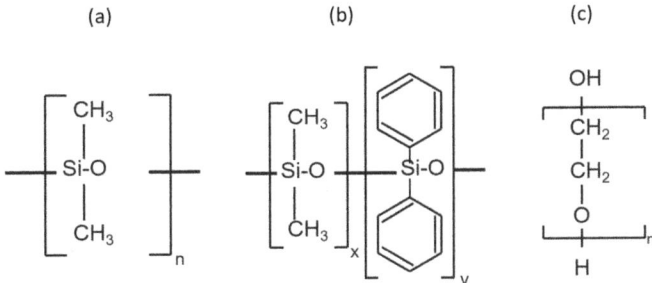

(a)   (b)   (c)

Figure 3.2 Types of column stationary phases: (a) polydimethylsiloxane; (b) dimethyl diphenyl siloxane; and (c) polyethylene glycol (PEG). Courtesy of Anthias Consulting.

phenols; and BASE or AMINE columns that are base deactivated PEG (or PDMS) for amines and basic compounds. Chiral columns are used for the separation of enantiomeric analytes with variations of β or γ cyclodextrin doped at different concentrations, usually onto 14% cyanopropylphenyl PDMS columns; these columns are less stable and should only be used for the separation of enantiomers. Ionic liquids are a completely new class of GC column phases and use solvents consisting entirely of ions. Their viscosity can be varied and they have unusual and highly tuneable selectivities, which are useful for some applications.

The type of stationary phase selected for the application is dependent on the chemistry of the analytes, for example their functional groups and their volatility. To separate analytes in the sample, their volatility and/or their interaction with the stationary phase must differ. A very good rule of thumb is that 'like separates like' or 'polar separates polar'. Start by considering the molecules (both the analytes and matrix) to be separated:

- What are their molecular masses and the range of volatilities?
- What types of functional groups do they have?
- What types of interactions could they undergo?

The temperature required to elute all analytes from the analytical column must always be a factor when selecting the phase type, as generally the more polar the analytical column, the lower the maximum temperature. The maximum temperature of the column must not be exceeded, otherwise the stationary phase will be damaged causing excess column bleed and higher baselines, greater activity within the column and generally poor peaks shapes, resolution and sensitivity.

The phase ratio is the ratio between the column internal diameter and the stationary phase thickness. Columns with lower phase ratios have a greater retention of analytes within the analytical column and therefore longer retention times; conversely, higher phase ratios have a lower retention and result in shorter retention times. Very volatile (low molecular weight) analytes are separated using columns with lower phase ratios which means that they have thicker stationary phases to trap the analytes. Less volatile (higher molecular weight) analytes are separated using thinner stationary phases, which allow a faster mass transfer of the analyte between the mobile and stationary phases, resulting in sharper peaks and better resolution.

3.3 Theoretical Background and Definition of Chromatographic Terms

In all chromatographic separations, solutes are partitioned between the mobile and stationary phases, this continuous process results in the separation of the solute components.

In gas chromatography the distribution into and out of the stationary phase of analytes is an equilibrium process and can be described as shown in eqn (3.1):

$$A_{gas} \rightleftharpoons A_{stationary} \qquad (3.1)$$

The equilibrium involved can be quantitatively described using a temperature dependent constant, the partition coefficient, K (eqn (3.2)).

$$K = \frac{C_s}{C_m} \qquad (3.2)$$

where C_s and C_m are the analytical concentrations of the components in the stationary and mobile phases respectively.

The mobile phase carries (elutes) the solute down the column (which holds the stationary phase) in a series of transfers between the mobile stationary phase. As solute movement can only occur in the mobile phase, the average rate at which a solute migrates depends on the fraction of time it spends in the stationary phase. All molecules spend the same average time in the mobile phase. Differences in the time spent in the stationary phase cause the components of a mixture to separate into bands along the length of the column. When a detector that responds to the solute is placed at the end of the column, the resulting response is plotted as a function of time and a chromatogram is obtained. This plot provides useful and important qualitative and quantitative information about the sample and the overall quality of separation.

A chromatogram (Figure 3.3) is a record of the concentration or mass profile of the sample components as a function of the movement of the mobile phase and the information it contains includes qualitative identification, quantitative assessment of the relative concentration of each sample component, an indication of the sample complexity and an indication of the column performance.

3.3.1 Retention on the Column

t_0 is the time taken for a solute that is not retained on the stationary phase (*i.e.* no interaction) to elute from the column. The un-retained solute therefore spends no time in the stationary phase and hence

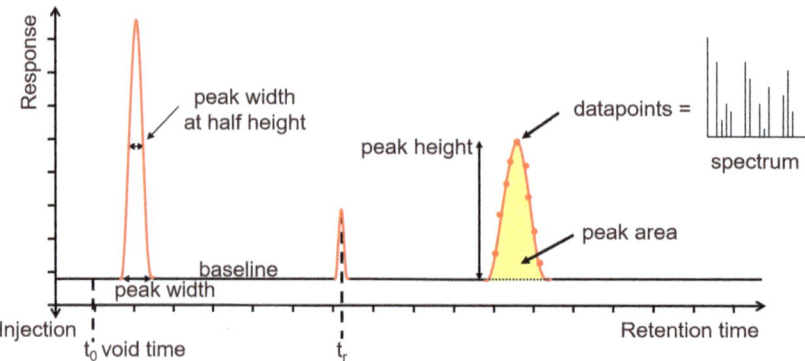

Figure 3.3   A chromatogram plot. Courtesy of Anthias Consulting Ltd.

has a partition coefficient of zero. t_0 is called the column dead time or the hold-up time. All solutes spend an equivalent time in the mobile phase, that is t_0. Retention is the capacity of a column (with a stationary phase) to retain a sample component. For solutes that are retained in the column, the retention time (t_r), that is the time between sample introduction (injection) and detector response, is greater than the column holdup time (t_0) using the time the solute spends in the stationary phase (t'_r) (eqn (3.3)).

$$t_r = t'_r + t_0 \tag{3.3}$$

Although retention is usually measured in time units, the volume of mobile phase required to elute a sample component (V_r) may be more accurately determined. Under normal experimental chromatographic conditions, liquids may be considered incompressible in liquid chromatography but this is not the case for gases in GC, and therefore GC elution volumes need correcting to a mean column pressure (eqn (3.4)). The retention volume will be a function of the mobile phase velocity.
  Thus:

$$V_r = F \times t_r \tag{3.4}$$

where F is the mobile phase flow rate and t_r is the retention time.
  The dead volume (V_0), the volume of mobile phase required to elute the un-retained solute (eqn (3.5)), is dependent on the column geometry and packing.

$$V_0 = F \times t_0 \tag{3.5}$$

The column inlet pressure is measured using a pressure gauge at the head of the column. If the flow rate through the column is measured using a soap film flow meter, it is necessary to correct for the vapor pressure of the soap film and the difference in temperature between the column and flow meter. Currently it is better to use a digital flow meter to record the carrier gas flow rate at the column outlet.

3.3.2 Capacity Factor

The capacity factor, k', also called the column partition ratio or retention factor is a measure of the time a solute spends in the stationary phase relative to the time spent in the mobile phase (eqn (3.6)).

$$k' = \frac{t_r - t_0}{t_0} \tag{3.6}$$

k' is widely used to evaluate the column performance and is a constant for a given compound in a particular chromatographic system. When the capacity factor for an analyte is less than one, the elution is so rapid that accurate determination of the retention time is very difficult. High retention factors (greater than 20) mean that elution takes a very long time. Ideally, the retention factor for an analyte should be between one and five.

3.3.3 Resolution

Resolution is an important consideration when developing separation methodologies and is a term used to describe the degree of separation of successive chromatographic bands or peaks. Resolution is achieved if the bands widen to a lesser extent than their maxima separate (eqn (3.7)).

$$R_s = \frac{(t_b - t_a)}{0.5(w_a + w_b)} \tag{3.7}$$

where R_s is the resolution of the two chromatographic peaks (Figure 3.4a). To increase the resolution of adjacent peaks requires either increasing the peak separation and/or reducing the peak widths. A resolution value of 1.0 refers to a 90% resolution while a value of 1.5 and above denotes baseline separation.

3.3.4 Selectivity Factor

The separation between two adjacent peaks (a and b) is also dependent on the relative retention characteristics of the two components, defined by their respective partition ratios (eqn (3.8)).

$$\alpha = \frac{k_b'}{k_a'} \tag{3.8}$$

where α is the selectivity factor (Figure 3.4b).

As the sample elutes through the column, its components will diffuse to form a Gaussian distribution under ideal conditions. This means that the parameters of the component band of the sample may be described using the Gaussian distribution equations.

$$w = 4\sigma$$

where w is the peak width and σ is the standard deviation of the peak (for a Gaussian distribution).

3.3.5 Band Broadening and Column Efficiency

To obtain optimal separations, sharp, symmetrical (Gaussian) chromatographic peaks must be obtained. This means that band broadening must be limited. Band broadening can be caused by many factors, including: poor introduction of the analyte onto the analytical column resulting in the analyte molecules spreading out; dead volumes within the system that are poorly swept by the carrier gas, the analyte molecules diffuse into these areas giving a slower progress; and irregular residence time in the stationary and mobile phases

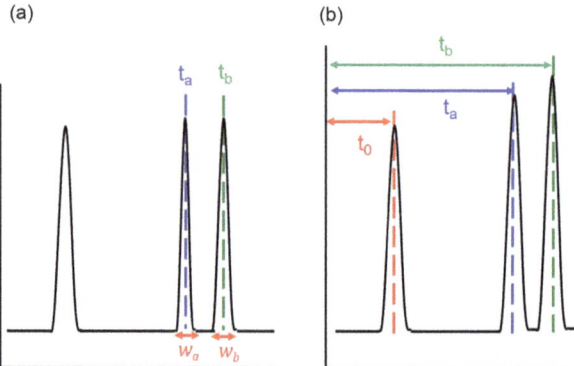

Figure 3.4   Measurements on the chromatogram used in the calculation of (a) the resolution, and (b) the selectivity and capacity factor. Courtesy of Anthias Consulting Ltd.

for each molecule of an analyte. Some may spend a greater than average length of time in the mobile or stationary phases, depending on whether they remain close to the mobile phase:stationary phase interface, or diffuse deeper into the stationary phase or into the centre of the analytical column. It is also beneficial to measure the efficiency of the column, which is a way of describing the narrowness of the peaks. Greater efficiency gives narrower peaks. Note that the peaks are Gaussian *because* the retention process is statistical.

The width of a solute band increases (broadens) as it elutes down the column because more time is allowed for band broadening processes to occur and therefore it is inversely proportional to the mobile phase velocity. The column efficiency, which is a measure of the band broadening, is measured by the number of theoretical plates (N) (eqn (3.9)). The concept of theoretical plates originates from the theory of distillation columns, based on the number of equilibria which may have occurred during a separation. It is important to remember that the plates do not really exist; they are a hangover from the days in which the separation of mixtures was performed using distillation columns, but this process can be seen as one movement of the analytes from the stationary phase into the mobile phase and back into the stationary phase again, as shown in Figure 3.5.

This model does however, serve as a useful way of measuring the column efficiency, either by stating the number of theoretical plates in a column, N (the more plates the better), or by stating the plate height, which is the height equivalent to a theoretical plate (the smaller the better).

$$N = 16\left(\frac{t_r^2}{w^2}\right) = \left(\frac{t_r^2}{\sigma^2}\right) \tag{3.9}$$

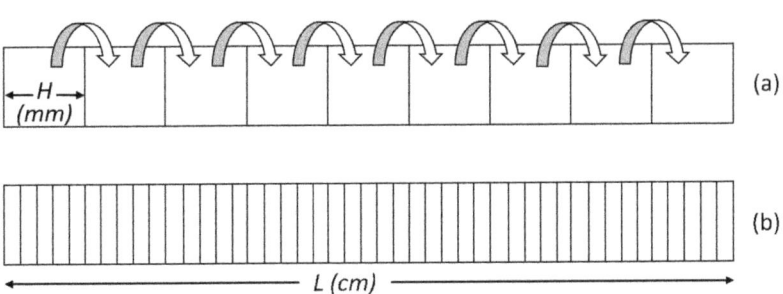

Figure 3.5 Model of theoretical plates in an analytical column of fixed length (cm): (a) a few theoretical plates and a large plate height; and (b) many theoretical plates and a small plate height. Courtesy of Anthias Consulting Ltd.

As can be seen from eqn (3.9), columns behave as if they have different numbers of plates for different solutes in a mixture. It can also be seen that the peak width is proportional to the retention time. This is important as you can often tell whether a peak 'belongs' in a particular chromatogram or is a later eluting peak from an earlier injection.

It is often easier and more precise to measure the peak width at some other position for example at the half peak height, $w_{1/2}$. The column efficiency is then related to $w_{1/2}$ using eqn (3.10):

$$N = 5.54\left(\frac{t_r{}^2}{w_{1/2}{}^2}\right) \tag{3.10}$$

The efficiency of a chromatographic column as a separation device improves as the number of equilibrations increases, that is the number of theoretical plates increases. The thickness of one theoretical plate, H or height equivalent to a theoretical plate (HETP) is calculated using eqn (3.11):

$$H = \left(\frac{L}{N}\right) \tag{3.11}$$

where L is the length of the separation column.

It is worth noting that although the plate height as a criterion of column efficiency is universally accepted, a plate as a physical entity in a column does not exist and the analogy of a chromatographic process to distillation is now largely irrelevant. For N and H to be meaningful in comparing two columns, it is essential that they be determined using the same compound.

3.3.6 Rate Theory and Band Broadening

We know that the separation process is an equilibrium process. A more scientifically rigorous description of the separation processes at work inside a column takes into account the time taken for the solute to equilibrate between the stationary and gas phases (unlike the plate model, which assumes that equilibration is infinitely fast). The resulting band shape of a chromatographic peak is therefore affected by the rate of elution. It is also affected by the different paths available to the solute molecules as they travel between and through the molecules of the stationary phase.

Several chemical and physical variables influence the rates at which bands in a column separate during elution and the amount of broadening for each of the bands. The effect of the mobile phase

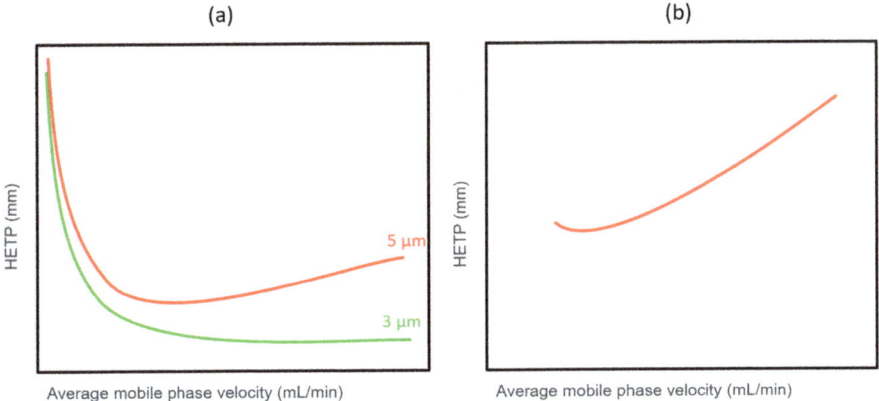

(a) (b)

HETP (mm)

Average mobile phase velocity (mL/min)

5 μm

3 μm

HETP (mm)

Average mobile phase velocity (mL/min)

Figure 3.6 Plots of plate height *versus* the mobile phase velocity for LC (a) and GC (b).

flow rate on the HETP for liquid chromatography (LC) and GC are compared (Figure 3.6).

When the HETP is small, the maximum efficiency will be realized for a given column length. The HETP for LC is larger compared to that of GC, LC uses packed columns and the particle size is important. Usually, the practical mobile flow rates are lower in LC (liquid mobile phases) compared with GC owing to the higher back pressures needed to pump liquids through the packed columns. Consequently, GC columns can be longer (up to 150 m) than LC columns.

Zone broadening is the increase in bandwidth as solutes pass through the column and it increases with an increase in the equilibrium steps.

$$H = A + \left(\frac{B}{u}\right) + \left(C_s + C_m\right)u \qquad (3.12)$$

This is the famous Van Deemter equation (eqn (3.12)), where u is the average linear velocity (eqn (3.13)). The average linear velocity is directly related to the speed of analysis, whereas the flow rate depends on the cross section of the column and the volume of the column occupied by the packing material:

$$u = \frac{L}{t_0} \qquad (3.13)$$

3.3.6.1 Eddy Diffusion Term

The eddy diffusion term (A) arises from the multitude of pathways which a molecule can use to find a path through a packed column, thus one molecule may take a different path to another, resulting in band broadening. Eddy diffusion is related to the particle diameter, geometry and the tightness of packing of the stationary phase. It is partly offset by ordinary diffusion, which results in molecules being transferred from one path to another. At low flow rates, a large number of transfers occur from one pathway to another. The rate at which each molecule moves down the column approaches that of the average. At high flow rates, the time is not sufficient for diffusion averaging to occur and band broadening due to eddy diffusion is observed. If using packed GC columns, eddy diffusion can be minimised by using homogenous columns with a narrow particle size and which do not have void spaces.

For open tubular columns, this term does not apply. As our discussion concerns GC-MS using capillary columns, eddy diffusion is absent, therefore the term A is zero and will not contribute to band broadening, this is then known as the Golay theory for open tubular columns.

3.3.6.2   Longitudinal Diffusion Term

The longitudinal diffusion term (B) is driven by the tendency of molecules to migrate from the concentrated centre of a band towards the more dilute outer regions. In LC diffusion is small compared with GC as diffusion rates in a liquid are smaller than those in gases. The rate of diffusion of an analyte is inversely proportional to the linear flow rate of the mobile phase. Longitudinal diffusion can be minimised by eluting analytes from the column as quickly as possible. Therefore, as far as possible and depending on the application, minimise the column length, maximise the carrier gas flow, use a low diffusion coefficient carrier gas, avoid large internal volumes (for example column i.d. and fittings), ensure connections are leak-tight and use appropriate transfer line temperatures. A rule of thumb for the transfer line between the GC and MS is 40 °C below the final oven temperature.

3.3.6.3   Mass-transfer Term

The mass transfer term is represented as C. The mobile phase is too fast for a true equilibrium of the solute molecules to be achieved between the stationary and mobile phases. The C_s term results from the resistance to mass transfer at the solute/stationary phase border. The C_m term results from mass transfer resistance in the mobile phase. Solute transfer between the phases is not instantaneous, leading to broadening at both ends of the band.

Taking into account all of the contributions to band broadening, an optimum velocity exists at which the plate height is at the minimum and the separation efficiency is maximised.

Band broadening from the mass-transfer term can be minimised by: using a thin stationary phase film to support faster movement between the phases; using smaller i.d. columns; the use of temperature programming to ensure reproducibility; and by ensuring the linear flow is not too fast.

3.4 How Do I Select the Best Mobile Phase to Give Me the Best Efficiency?

We have already seen the relationship between band broadening and the efficiency of separation.

Figure 3.7 shows a plot of the height equivalent to a theoretical plate of a column *versus* the mobile phase velocity of the carrier gas. This is known as a Van Deemter plot and it shows the contributions from the A, B and C terms of the Van Deemter equation (eqn (3.14)).

$$\text{HETP} = A + \left[\frac{B}{u}\right] + (C_s + C_m)u \tag{3.14}$$

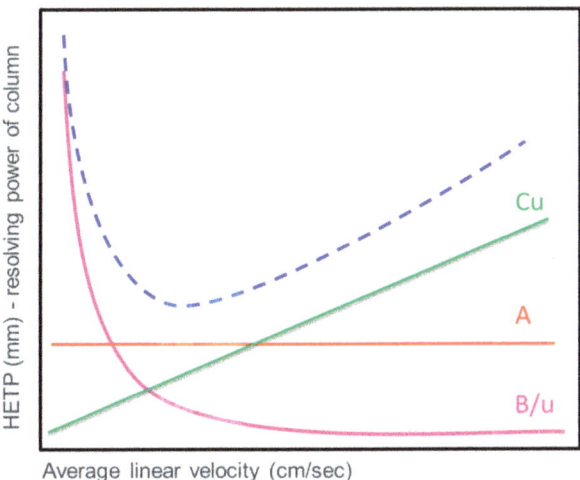

Average linear velocity (cm/sec)

Figure 3.7 Contributions of each of the eddy, longitudinal diffusion and mass-transfer parameters to the overall band broadening. Courtesy of Anthias Consulting Ltd..

The optimum mobile phase velocity is obtained by taking the first differential of the equation for HETP with respect to the mobile phase velocity (u) and equating it to zero.

Thus:

$$U_{opt} = \left[\frac{B}{(C_s + C_m)} \right]^{0.5} \tag{3.15}$$

And therefore:

$$\text{HETP}_{min} = 2[B(C_s + C_m)]^{0.5} \tag{3.16}$$

This means that the highest column efficiency, which is realised at the optimum mobile phase velocity is related to the longitudinal diffusion of the solute molecules in the mobile phase (eqn (3.15) and (3.16)). The type of the carrier gas, whether it is hydrogen, helium or nitrogen will therefore directly influence the separation efficiency.

It can be seen in Figure 3.8 that as the molecular weight of the gas decreases, the optimum gas velocity increases, and also the shape of the curve flattens. We are therefore able to use higher gas velocities for hydrogen carrier gas without compromising the efficiency compared with the use of nitrogen gas and also small changes in the range of velocity will not significantly affect the practical efficiency.

However, nitrogen is not frequently used in capillary column oven temperature programmed GC, owing to its high viscosity and low diffusivity, it also not used with GC-MS. Hydrogen is the best carrier gas in terms of the lowest viscosity and highest diffusivity, resulting in fast,

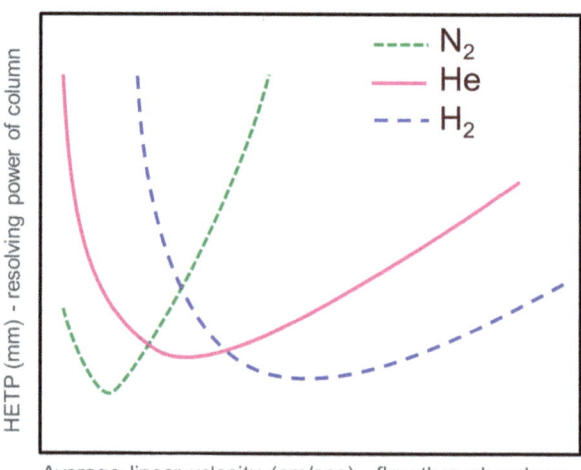

Figure 3.8   Van Deemter curves for different carrier gases. Courtesy of Anthias Consulting Ltd.

temperature-programmed, high efficiency separations, however, it is highly flammable and is reactive. Therefore, it is less frequently used in GC-MS applications in which the sensitivity can be four times lower than with helium, or in the analysis of unknowns in which reactions could occur and if there are safety concerns. This means that for sample analysis with GC-MS, helium is by far the most popular carrier gas enabling inert, high resolution, temperature-programmed separations.

3.5 How Do I Reduce Retention Time?

There are three options available to reduce the retention time (t_r):

- Reduce L (column length).
- Reduce k (retention factor) by increasing the temperature.
- Increase u (carrier gas linear velocity).

Using eqn (3.17) and (3.18) we can define the GC retention time:

$$t_r = \frac{L(k+1)}{u} \tag{3.17}$$

$$t'_r = \left(\frac{C_s}{C_m}\right)\left(\frac{2d_f}{r}\right)\left(\frac{L}{u}\right) \tag{3.18}$$

where:

- t'_r is the adjusted retention time, $t'_r = t_r - t_0$.
- d_f is the film thickness of the stationary phase.
- C_s/C_m is the distribution constant, which is related to k.

The higher the value of k, the greater the concentration of solute molecules in the stationary phase and the longer the retention time. Changing the temperature changes the distribution coefficient, thus temperature programming in GC is effectively programming the distribution constant. Increasing the film thickness increases the retention times and also gives increased capacity for sample handling (Table 3.1). The amount of sample introduced into the column may influence the peak shapes, particularly if overloading occurs.

Increasing the column radius has the effect of reducing the retention times as the solute molecules spend more time in the mobile phase, while increasing the length of the column increases the retention times.

Table 3.1 Variation of sample capacity as the column diameter and stationary phase film thickness are increased.

Column i.d. (mm)	Approximate sample capacity (ng)			
	Film thickness (µm)			
	0.1	0.25	0.5	1.0
0.1	10	30–40	50–70	100–200
0.18	20–30	60–80	100–150	250–350
0.25	30–40	125–175	175–250	400–500
0.32	50–70	200–250	250–350	600–800
0.53	100–120	400–500	500–700	1000–1500

A reduction in the retention time results in more analysis being performed per given time, increasing the productivity of the laboratory.

The choice of mobile phase is usually between helium and hydrogen. Currently more laboratories are using hydrogen as a carrier gas, although it carries the risk of explosion. However, the fitting of a hydrogen leak detector reduces this risk, and the use of a hydrogen generator significantly reduces the laboratory costs as it is not possible to locally produce helium gas. It should be noted that commercially available MS libraries, including National Institute of Standards and Technology (NIST) libraries have been generated using helium carrier gas and the use of hydrogen may change the mass spectra and invalidate the use of routine library searching.

In capillary gas chromatography, there is a pressure drop along the column owing to the compressibility of the gas mobile phase. The stationary phase mass transfer term is influenced by the column pressure drop at each and every position along the column.

The mass transfer in the mobile phase contribution increases with the increasing column diameter and lower capacity values. Mass transfer at the stationary phase increases in importance as the film thickness increases. For thin film columns (df < 0.25 µ), the contribution of the C_s term to band broadening is low compared with the longitudinal diffusion contribution. There is a strong dependence of the C_s term contribution to band broadening on capacity factors and diffusion coefficients in the stationary phase. In general terms, thick-film polar capillary columns will offer a reduced separation efficiency compared to similar apolar columns, and the efficiency in both cases also decreases with the increasing film thickness. Increasing the stationary film thickness increases the sample capacity of the column (Table 3.2).

Table 3.2 Influence of column i.d. on column efficiency and sample capacity.[a]

Internal diameter (mm)	Efficiency: Plates/ Meter ($N\ m^{-1}$)	Efficiency: Total plates (N)	Capacity each analyte (ng)
0.53	1300	39 000	1000–2000
0.32	2300	69 000	400–500
0.25	2925	87 750	50–100
0.20	3650	109 500	<50
0.18	4050	121 500	<50
0.10	7300	219 000	<10

[a]Theoretical values for 30 m columns, calculated at $k = 6.00$ and an 85% coating efficiency.

Band broadening in open tubular columns has been discussed in detail using the Golay equation, taking into consideration the following, which refer to capillary columns:

- carrier gas diffusion coefficient at the column outlet pressure
- carrier gas velocity at the column outlet;
- stationary phase film thickness;
- ratio of the column inlet to the column outlet pressure.

3.6 How Do I Evaluate the Separation of Closely Eluting Bands in a Chromatogram?

The separation of the bands of a mixture as they elute from a chromatographic column is the key principle of chromatography. Broad bands are likely to resolve poorly compared with sharp narrow bands (Figure 3.9).

The selectivity factor α is a measure of the relative peak separation (eqn (3.19)):

$$\alpha = \frac{(k_a')}{(k_b')} \tag{3.19}$$

such that $\alpha \geq 1$.

The degree of separation between two peaks is given by the resolution (eqn (3.20))

$$R_s = 2\frac{(\Delta Z)}{(w_a + w_b)} \tag{3.20}$$

A resolution greater than or equal to 1.5 means that the peaks are baseline separated:

$$R_s = \left(\frac{\sqrt{N}}{4}\right)\left(\frac{k}{k+1}\right)\left(\frac{\alpha - 1}{\alpha}\right) \tag{3.21}$$

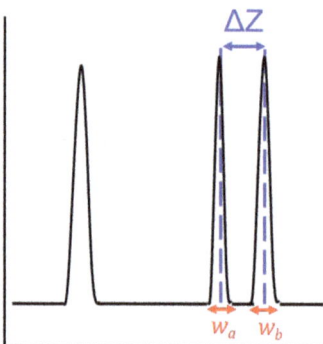

Figure 3.9   Resolution between closely eluting bands. Courtesy of Anthias Consulting Ltd.

From eqn (3.21) shown above, the resolution only increases in proportion to the square root of the efficiency. This means that doubling the length of the column, which doubles the number of theoretical plates increases the resolution only by the square root of 2. The efficiency of a column increases with a decrease in the column i.d.

3.7   How Do I Optimise Column Performance?

The optimal conditions will give a baseline separation for all eluting peaks in the minimum amount of time. Below are some suggestions for optimising chromatography analysis.

- Increase the total number of theoretical plates (efficiency) by lengthening the column. This negatively impacts the time required for the separation.
- To increase N without increasing t, reduce HETP. Use columns with a thinner phase thickness and/or a smaller column diameter.
- An increase in the capacity ratio, k', enhances the resolution but also increases the separation time.
- In GC, k' values are improved by increasing the temperature of separation.

When the selectivity factor (α) is approximately equal to 1, optimizing the k' and increasing N will not improve the resolution. To increase the selectivity, the options available in the GC are to change the stationary phase (use a column with a stationary phase of different polarity) or change the temperature. Usually changing the temperature and temperature programming are attempted to improve the separation before changing the column.

Further Reading Materials

1. A. Braithwaite and F. J. Smith, *Chromatographic Methods*, Blackie Academic and Professional, London, 5th edn, 1996.
2. H.-J. Hubschmann, *Handbook of GC/MS*, Wiley – VCH Verlag GmbH & Co, Weinheim, 2nd edn, 2009.
3. A. Fowlis Ian, *Gas Chromatography*, John Wiley & Sons, Chichester, 2nd edn, 1999.
4. C. F. Poole and S. K. Poole, *Chromatography Today*, Elsevier Science, Amsterdam, 5th edn, 1997.

4 How Do I Detect My Analytes?

4.1 What Types of GC Detectors Are There?

There have been over 60 detectors designed for GC instrumentation, of which around 15 are commonly used and commercially available for analysis.

Although this book focusses on GC-MS, in this chapter, we will provide a brief overview of the few GC detectors available and the important parameters that should be considered if you wish to split your column effluent between a MS and a GC detector. GC detectors can have advantages over MS for some applications, for example the quantitation of hydrocarbons. Some of the commonly available detectors are:

- Flame ionisation detector (FID)
- Electron capture detector (ECD)
- Thermal conductivity detector (TCD)
- Flame photometric detector (FPD)
- Nitrogen phosphorus detector (NPD)
- Barrier discharge ionisation detector (BDID)
- Photo ionisation detector (PID)
- Helium ionisation detector (HID) and discharge ionisation detector (DID)
- Atomic emission detector (AED)
- Sulphur chemiluminescence detector (SCD)
- Nitrogen chemiluminescence detector (NCD)
- Electrolytic conductivity detector (ELCD)

Gas Chromatography-Mass Spectrometry: How Do I Get the Best Results?
By Diane Turner, Mathias Schäfer, Steven Lancaster, Imran Janmohamed, Anthony Gachanja and Jason Creasey
© Diane C. Turner, Mathias Schäfer, Steven Lancaster, Imran Janmohamed, Anthony Gachanja and Jason Creasey 2020
Published by the Royal Society of Chemistry, www.rsc.org

- Thermal energy analyser (TEA)
- Infra-red detector (IRD)
- Vacuum ultra-violet (VUV)

For the selection of an ideal detector, you will need to consider if the detector will:

- Have adequate sensitivity for your target analytes.
- Give a similar response to all analytes, or be selective to one or more classes of analytes.
- Be non-destructive of the sample if you wish to put multiple detectors in series.
- Have a dynamic range that is suitable for your application, preferably with a linear response to the target analytes over several orders of magnitude.
- Give a quick response time that is independent of the flow rate, this is particularly important for narrow peaks especially if using GC×GC.
- Give good stability and reproducibility at the instrument location.
- Be easy to use, set-up and maintain by the instrument operators.

However, no detector exhibits all of these qualities and you will have to choose one that is best for the application!
Detectors can be of two types:

- *Universal*: will respond to most organic compounds and some will respond to inorganic compounds too. Universal detectors are good for screening unknown mixtures and for situations in which it is important to ensure all solutes are detected and quantified. Examples include FID, TCD and MS in full-scan acquisition mode.
- *Selective*: responds to a specific type or class of compound. Selective detectors usually show enhanced sensitivity to certain compounds and can make complex chromatograms simple by only detecting certain classes of compound. Examples include SCD which only responds to sulphur; ECD which only responds to compounds capable of capturing an electron such as halogen atoms; and MS in selected ion monitoring (SIM) or MS/MS acquisition mode.

As can be seen in Figure 4.1, if we inject a sample into a GC-FID, you will obtain a particular response that represents the characteristic hydrocarbon moiety, that is it can see anything with a C–H bond. If we inject the same sample into a GC-ECD, you will observe a different

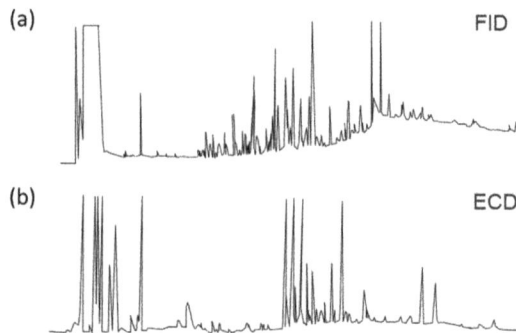

Figure 4.1 A chromatogram from a GC-FID (a) and a chromatogram from a GC-ECD (b) of the same sample showing the different charaterictic responses from the detector.

chromatographic pattern as the detector will be responding to a different characteristic such as the electronegativity, that is all of the peaks seen captured an electron.

Detectors can also be characterised as being concentration or mass flow sensitive. Concentration sensitive detectors are usually non-destructive detectors that respond to the concentration of the analyte in the detector flow cell or chamber, for example TCD, ECD and PID. Mass flow sensitive detectors are usually destructive detectors that respond to the amount of analyte in the detector at any time, irrespective of the volume of the carrier gas, for example FID, PFPD, NPD, ELCD, AED or a mass selective detector (MSD)

4.2   What Is a FID?

The FID is one of the most widely used detectors available today. It is a destructive, mass sensing, non-selective detector that provides information on the retention time and response only. This detector is a "carbon–hydrogen bond" counter, as the detector response is proportional to the carbon number of the hydrocarbon molecules. The compounds and functional groups that give little/no response include:

- carbonyls, COH and COOH;
- alcohols, halogens and amines;
- non-combustible gases, H_2O, CO_2, CO, SO_2, NO_x, N_2, O_2, NH_3 and rare gases.

Hence, it is a useful detector for analysing samples that are contaminated with water, nitrogen oxides and sulphur oxides, as it is will not respond to these.

(a)

(b)

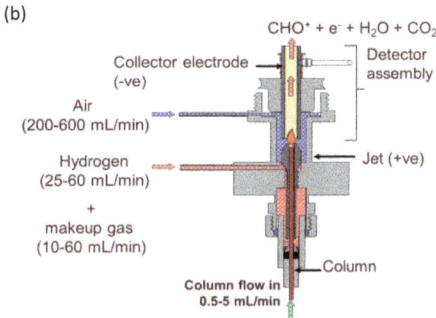

Figure 4.2 (a) Picture of an Agilent FID and (b) a schematic diagram of a typical FID detector. Courtesy of Anthias Consulting Ltd.

4.2.1 How Does a FID Work?

In this detector, the column effluent is mixed with hydrogen and a make-up gas (for capillary systems) before exiting *via* a small orifice (or jet-tip) which is surrounded by a high flow of air (see Figure 4.2). Hydrogen is combustible in air and is lit *via* a remote glow plug. The organic compounds burn producing cations (CHO^+), electrons, water and CO_2. The cations are collected and produce a signal, which is measured by the collector electrode.

The high sensitivity and wide linear range of the FID which gives a molar response for carbon-containing compounds make it very popular in organic analysis. It is cheap, robust, low maintenance and easy to operate, it can detect the majority of organic compounds and is a good detector to use as a pre-screening detector before conducting analysis using highly sensitive and more expensive techniques. However, it needs at least two detector gases (hydrogen and air), responds to almost everything (the hydrocarbon background from oil, cars, *etc.*), it destroys the sample and it does not give much information about the identity of the analyte.

4.2.2 What Parameters Do I Need to Optimise When Using a FID?

There are several parameters that need to be considered when optimising the sensitivity of the FID. These include:

- Detector (and carrier) gases: these need to be free of moisture and hydrocarbons – use gas filters.
- Optimising hydrogen flow: typical 30–40 mL min^{-1} (caution, it is not linear!), this will need to be reduced if H_2 is used as the carrier gas.
- Increasing air flow (to a point): typically 300–400 mL min^{-1}.

- Selecting the best make-up gas: N_2 > Ar > He (the same as for the carrier gas).
- Optimising the make-up gas to increase flow through the jet: typical column plus makeup gas flows are 30 mL min^{-1}.
- Temperature: this should be no higher than the column maximum isothermal temperature, not so high as to degrade analytes, and not so cold as to cause condensation of water or analytes or to increase longitudinal diffusion. As a rule of thumb, it should be equal to or down to 20–30 °C lower than the final oven temperature.

4.3   What Is an ECD?

The ECD is a selective ionisation detector which measures the electrical conductivity of the column effluent resulting from ionising radiation from a radioactive ion source. It only responds to analytes capable of capturing an electron, these include:

- Halides
- Nitrates and nitriles
- Peroxides, anhydrides and oxygen
- Organometallics
- Conjugated double bonds

4.3.1   How Do an ECD and µECD Work?

The radioactive ^{63}Ni isotope plated onto the inner surface of the cell body emits beta particles (Figure 4.3). These negatively charged particles collide with the nitrogen make-up gas molecules and produce higher energy electrons and excited nitrogen. These electrons produce a high, constant current between the collector anode (*e.g.* the detector body) and the cathode (usually a cylindrical electrode in the centre or top of the detector) with a potential difference applied.

An electronegative analyte, such as a halogenated compound, elutes from the GC column and captures some of the electrons, this results in the reduction of the cell current. The extent of electron absorption, and hence the reduction in the standing current, is proportional to the concentration of the analyte. In the µECD, a wave pulse is applied to the detector, the modulation is varied to provide a constant voltage, which improves the linear range and reduces tailing. This pulse frequency is recorded as a chromatogram. The µECD anode responds only to electrons generated by the nitrogen make-up gas and not the heavier slow moving analyte ions which are purged out with the carrier gas.

(a)   (b)

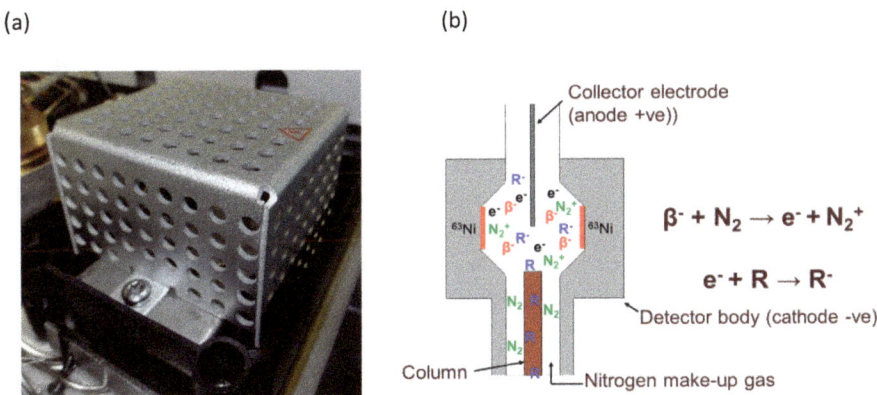

Figure 4.3   (a) Picture of an Agilent μECD, and (b) a schematic diagram of a typical ECD including reactions. Courtesy of Anthias Consulting Ltd.

The very high sensitivity of the ECD of around 50 fg for an electron capturing analyte (6 fg mL^{-1} for lindane) and the selectivity to certain compound classes make it one of the most popular selective detectors. It is low maintenance and easy to operate. However, it requires time to stabilise which can be between 2 to 7 hours (up to 2 weeks for a contaminated detector); it can become saturated from high concentration analytes; it cannot be used with halogenated solvents such as dichloromethane (DCM); it can be de-sensitised by air leaks as it is very sensitive to oxygen; it requires high electron affinity and electronegative analyte atoms or bonds to respond; and it does not give much information about analyte identification apart from the retention time and response.

4.3.2 What Parameters Do I Need to Optimise When Using an ECD?

There are a few parameters that need to be considered when optimising the sensitivity of an ECD. These include:

- Nitrogen make-up (and carrier) gases: these need to be free of moisture and oxygen – use gas filters. Trace levels of oxygen and water, both of which are electronegative, can produce a noisy and high baseline. It is recommended to use ultra-pure gases to ensure an optimum performance.
- Carrier gas type: nitrogen gives better sensitivity than helium; argon/methane mixtures give a better dynamic range.
- Make-up gas flow: typically 30–60 mL min^{-1} of carrier + make-up gases, this should be optimised.

- Column ferrules: graphite ferrules are porous to oxygen, therefore, if possible (manufacturer dependent), use graphitised vespel or metal ferrules to connect both ends of the column in the GC, especially to the detector.
- Temperature: no higher than the column maximum isothermal temperature, not so high as to degrade the analytes, not so cold as to cause condensation of water or the analytes or to increase the longitudinal diffusion. As a rule of thumb, it should be equal to or down to 20–30 °C lower than the final oven temperature.

4.4   What Is a Thermal Conductivity Detector?

The TCD is a non-destructive, concentration sensing, near-universal detector which measures changes in the thermal conductivity of the column effluent compared to a reference flow of carrier gas.

4.4.1   How Does a TCD Work?

A heated filament, made of rhenium, tungsten, gold or a semi-conducting thermistor, has a temperature that is dependent on the type of carrier gas that surrounds it and it is cooled by this flow of pure carrier gas producing a constant reference reading. A separate filament may be used as a Wheatstone bridge, or the flow passing the filament is switched (five times per second) between the reference flow and the column flow with a make-up gas (MUG) added. Analytes eluting from the GC column will have a thermal conductivity that is significantly different from the common carrier gases helium and hydrogen. To maintain the filament temperature when an analyte elutes, a current is applied, this is related to the analyte concentration and is recorded as a chromatogram (Figure 4.4).

The TCD is not a very sensitive detector but it is simple, cheap and low maintenance; responsive to both organic and inorganic analytes; non-destructive to the sample and therefore can be hyphenated with other detectors, for example a FID.

4.4.2   What Parameters Do I Need to Optimise When Using a TCD?

There are a few parameters that need to be considered when optimising the sensitivity of a TCD. These include:

- Flows: it is crucial to optimise the reference and sample flows to obtain the best sensitivity. The recommended flows (manufacturer and application dependent) are:

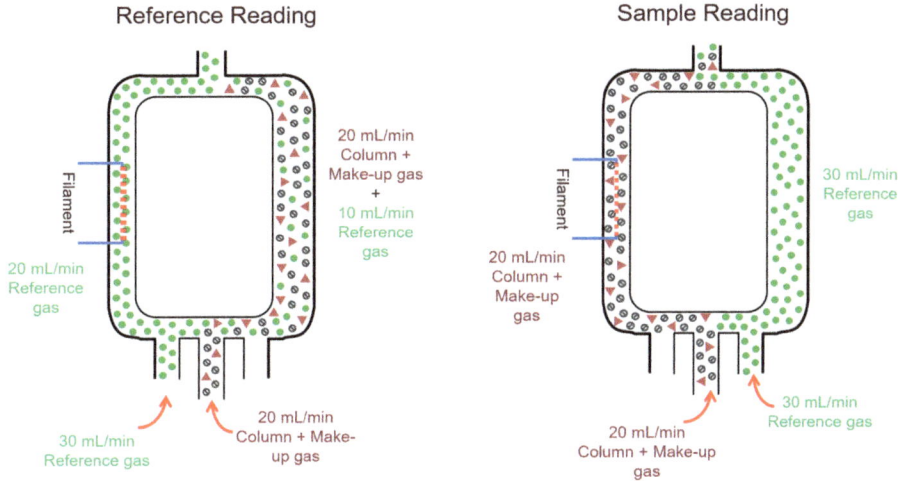

Figure 4.4   Schematic of a typical TCD. Courtesy of Anthias Consulting Ltd.

o packed column flow: 10–60 mL min^{-1};
o capillary column flow: 1–5 mL min^{-1};
o reference flow: 15–60 mL min^{-1};
o make-up flow: 5–15 mL min^{-1} (capillary), 2–3 mL min^{-1} (packed).
• Detector temperature: for optimum results, it is important that the detector temperature is 30–50 °C higher than the highest oven temperature, be wary of not exceeding the maximum iso-thermal column temperature.

4.5   Why Would I Choose a GC-MS Over Other GC Detectors?

The different types of detectors can respond to specific elements, bonds or functional groups in a molecule, or the physical properties such as the electronegativity or thermal conductivity.

The advantages of GC-MS compared to other GC detectors are:

• A GC-MS produces a third dimension of data which can be pertinent to the identification of unknown compounds *via* spectral interpretation or library searching.
• Can give more certainty in quantitative target analysis in which you have a retention time and mass spectral confirmation.
• Chromatographic separation and the spectral separation of analytes: deconvolution of coeluting peaks.

However, GC-MS may be unnecessary if analytes are known and well separated. Universal detectors are good for seeing most organic

Table 4.1 GC detector types, what they respond to, their sensitivity and linear range.

Detector	Known as	Analytes/atoms /bonds	Sensitivity	Linear range
Flame ionisation detector	FID	C–H bonds	100 pg C	10^7
Pulsed flame photometric detector	PFPD	Sulphur/phosphorus /other elements	S & P fg	S: 10^3 P: 10^3
Photo ionisation detector	PID	Volatile organic compounds (VOCs)	25–50 pg	10^{5-7}
Vacuum ultra-violet	VUV	Functional groups, isomers	pg	10^4
Infra-red detector	IRD	Functional groups, isomers	ng	10^3
Sulphur/nitrogen chemiluminescence detector	SCD/NCD	Sulphur/nitrogen	pg	10^4
Thermal energy analyser	TEA	Nitrogen groups	pg	10^4
Atomic emission detector	AED	Heteroatoms and other elements	< pg	10^4
Thermal conductivity detector	TCD	Organic and inorganic	< ng	10^5
Nitrogen phosphorus detector	NPD	Nitrogen/ phosphorus	50–500 fg	10^5
Barrier discharge ionisation detector	BID	Organic	Low pg	10^5
Helium/discharge ionisation detector	HID/DID	Gases	0.1 ppm	10^2
Electron capture detector	ECD	Electron capturing, halides	< 50 fg	10^4
Electrolytic conductivity detector	ELCD	Halogens, sulphur, nitrogen	5 ppb	10^3

compounds in a sample, whereas selective detectors can reduce matrix interferences and improve detection limits. As shown in Table 4.1, some detectors give a similar response to all analytes, others are selective to one or more classes of analytes. Some detectors can be more selective and specific than GC-MS, whilst other can be more sensitive for certain analytes, for example ECD, NPD, SCD and NCD. Some have a greater linear range or might give more meaningful data, such as a FID for the quantitation of hydrocarbons.

It is vitally important to select the most appropriate detector for the analytical application that you wish to consider, and it may not always need to be an MS.

4.6   How Do I Connect My Analytical Column to Two or More Detectors?

Splitting the analytical column effluent between two detectors can be very useful, for example to see different classes of compounds with a better sensitivity using two different selective detectors (Figure 4.5),

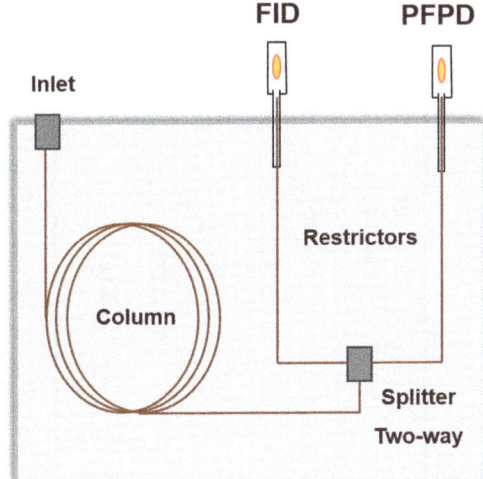

Figure 4.5   Example of a two-way splitter to split the column effluent between a FID and PFPD. Courtesty of Anthias Consulting Ltd.

or to use one detector such as FID for quantitation and MS for identification.

There several types of splitters to split the effluent between the detectors:

- press-fit connector;
- union;
- splitter plate which could be purged and can be 2-way or 3-way.

Considerations:

- Splitter should have no dead volume, otherwise the peak shape will suffer.
- Ensure there are no leaks: columns should be trimmed squarely and the manufacturer's instructions of how to install it should be followed.
- Splitter should be of low thermal mass: enabling rapid heating and cooling in the GC oven with no cold or hot spots.

Controlling the internal diameter and lengths of the restrictors connecting the splitter to each detector means that the split ratios between the detectors can be altered. The pressure of the detector also needs to be taken into consideration, for example a FID is an atmospheric pressure detector, whereas an MS has a vacuum. There are calculators available to determine the lengths and internal diameter of these restrictors, some examples are given in Table 4.2.

Table 4.2 Restrictor length and the internal diameters used to control split ratios.

Configuration	Detector 1	Detector 2	Detector 2/ Detector 1 split ratio	Diameter restrictor detector 1 (mm id)	Length restrictor detector 1 (m)	Diameter restrictor detector 2 (mm id)	Length restrictor detector 2 (m)
1	Atmospheric	Atmospheric	1	0.25	0.544	0.25	0.544
2	Atmospheric	Atmospheric	5	0.18	0.418	0.25	0.311
3	Atmospheric	MSD diffusion or standard turbo	1	0.18	1.060	0.18	2.890
4	Atmospheric	MSD diffusion or standard turbo	2	0.18	2.130	0.18	2.890
5	Atmospheric	MSD diffusion or standard turbo	5	0.10	0.507	0.18	2.890

5 Mass Analysis

5.1 A Historical Perspective: Why Do I Need Ions for MS Analysis?

Mass spectrometry (MS) evolved from the study of electrical interactions with gases in the mid 1800s by Sir Joseph John Thomson. This early work resulted in the discovery of cathode rays (later found to be electrons), and anode rays (positive ions). Thomson (18 December 1856–30 August 1940) was an English physicist and was appointed to the Cavendish Chair of Experimental Physics at Cambridge University in 1884. It was found that anode rays could be 'separated' in a magnetic field as performed in his famous parabola spectrograph (Figure 5.1). Through this work he discovered the electron, suggesting that it was more than 1000 times less massive than a hydrogen atom. He also suggested that these subatomic particles were always the same, irrespective of what type of atom they came from ref. 7. In subsequent experiments with cathode rays, Thomson found that they could be deflected by both electric and magnetic fields and he obtained experimental evidence for the *mass-to-charge ratio* (m/z) of the electron.[1] The first sector field mass spectrometer was the result of this work. The principles of ion separation in a magnetic field are illustrated by the three fingers rule as shown in Figure 5.2.

Gas Chromatography-Mass Spectrometry: How Do I Get the Best Results?
By Diane Turner, Mathias Schäfer, Steven Lancaster, Imran Janmohamed, Anthony Gachanja and Jason Creasey
© Diane C. Turner, Mathias Schäfer, Steven Lancaster, Imran Janmohamed, Anthony Gachanja and Jason Creasey 2020
Published by the Royal Society of Chemistry, www.rsc.org

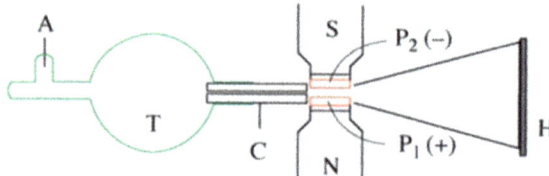

Figure 5.1   Schematic diagram of Thomson's parabola-spectrograph, 1897. Ions are
generated by a discharge that occurs in the spherical tube, T, between
the anode, A, and the cathode, C, which is a fine tube. The ions are
deflected by the magnetic field of an electromagnet (magnetic poles N
and S) and by an electric field applied to the poles, P_1 and P_2. The ions
finally strike a photoplate, H, in which they produce a fluorescent spot.
The *m/z* ratio is deduced from the position of the traces on the photoplate in relation to the experimental settings. His work demonstrated
that ions of different mass and kinetic energy can be analysed in a mass
spectrometer. These experiments delivered the final proof of the atomic
theory of matter. To distinguish between the two Ne isotopes, Thomson made use of the different behaviour of charged particles of varying
momentum and energy in an electromagnetic field. Reproduced from
ref. 2 with permission from the Royal Society of Chemistry.

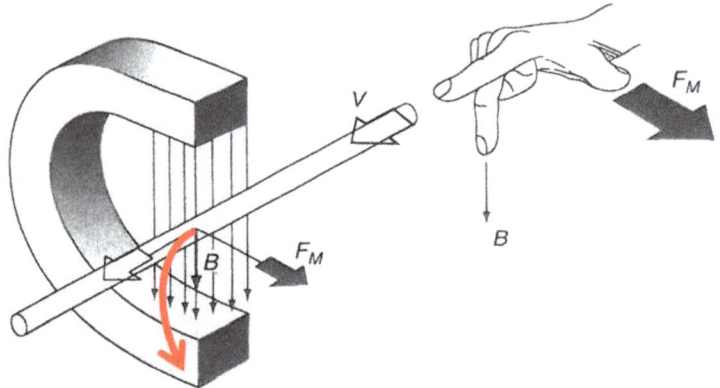

Figure 5.2   The magnetic field B has a direction that is perpendicular to the flight-
path of the accelerated ion with velocity *V*. In consequence a magnetic
force F_M, that is, the Lorentz-Force, is submitted to the ion that will lead
to a deflection (red arrow). The three fingers of the right hand illustrate
this relationship, that is the orientation conventions for vectors in three
dimensions. Adapted from ref. 6 with permission from John Wiley and
Sons, Copyright © 2007 John Wiley and Sons.

Hence, Thomson's seminal work demonstrated that mass spectrometric analysis delivers the *m/z* ratio and not just the mass (MS instruments are therefore not balances) and that charged particles such as
cations, anions or even electrons need to be manipulated (accelerated,

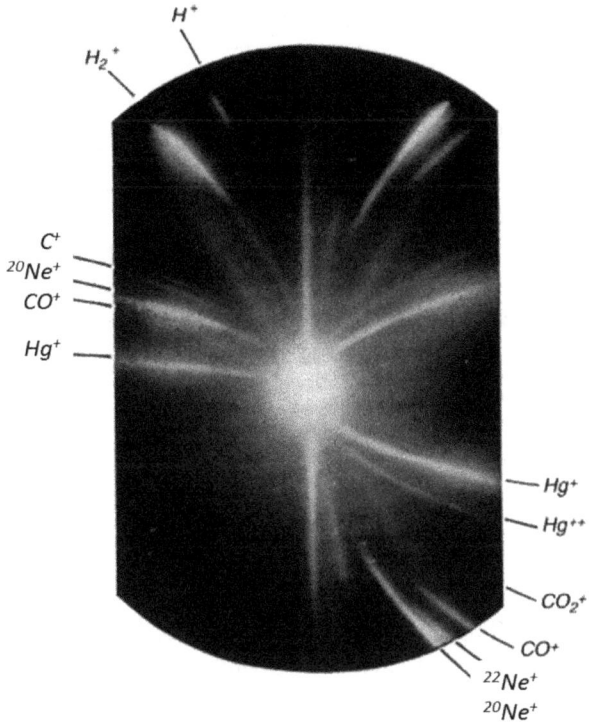

Figure 5.3 Photoplate from Thomson's parabola-spectrograph with neon.[3-5] Reproduced from ref. 3 with permission from Elsevier, Copyright 2006.

deflected *etc.*) with magnetic and/or electric fields (Figure 5.3). This discovery enabled Thomson to propose that electrons emerged from the atoms of gas in the cathode ray tubes, thus concluding that atoms were in fact divisible.

5.1.1 Are Mass Spectrometers Sophisticated Balances That Are Able to Determine the Mass of a Molecule?

This short historical overview demonstrates that MS is an important technology that has facilitated many very significant scientific advances. Of these advances in science the discovery of isotopes, the exact determination of molecular mass, the characterization of new elements, quantitative gas analysis, isotope labelling, rapid identification of trace pollutants and drugs and the characterization of the molecular structure are very significant.[7] Indeed, MS is perhaps the most widely used technique in the world today for the identification and quantitation of compounds.

5.2   Introduction to Mass Spectrometry

5.2.1   What Are the Basic Fundamentals Needed for Understanding MS?

Mass Spectrometry is the term used to describe a collection of techniques which separate ionic species by means of their *mass-to-charge* ratio (*m/z*). It is a powerful tool in the armoury of the analytical chemist and enables the identification of analytes, that is the qualitative analysis of clean compounds, as well as mixtures using GC-MS, and ultimately also the quantitation of analytes. Before going into the complexities of MS, this chapter will give a brief introduction to the terms used and the indispensable fundamentals that are important for understanding MS.

5.2.2   Mass

Mass is a measure of the amount of matter, in other words, it is the quantity of matter. The SI unit of mass is the kilogram (kg). For the purposes of this discussion, it is a fundamental property and unlike weight, does not depend on where it is measured. That is, for a given amount of substance, the mass is independent of gravity and (at non-relativistic velocities) velocity. Weight is dependent on the force acting on a body, and will therefore be different when experienced under different gravitational fields. We should not use the terms interchangeably.

5.2.3   Atoms of Elements and Isotopes

The ancient Greek philosopher, Democritus (460–370BC) first proposed the concept of the atom as an indivisible constituent of matter[8] and John Dalton (1766–1844), an English chemist developed the first modern atomic theory in 1808.[9] In general the atom is the smallest constituent of matter which has the properties of a chemical element.[10] However, it took another century until Rutherford conducted the famous gold foil experiment in 1911, establishing the model in which atoms of chemical elements consist of a solid nucleus with a characteristic number of positively charged protons (which is the *atomic number* of the respective element, and also determines its place in the *periodic table of elements*) and an equal amount of negatively charged electrons which are found outside of the nucleus and balance the positive charge of the nucleus.[11]

The nucleus of an atom consists of protons and, of neutrons (with the exception of the hydrogen atom ^{1}H, which has only a proton in the nucleus). The *atomic number* is conventionally written as a subscript preceding the elemental symbol, for example $_{11}$Na for sodium and $_6$C for carbon. Atoms of one element that differ in the number of neutrons in their nuclei differ in their mass. Those nuclides are called *isotopes* (Greek: *isos topos – same place* in the *periodic table of elements*, see below). The sum of the protons and neutron equals the *mass number* of any sort of atom, that is any nuclide. The *mass number* is written as a superscript preceding the elemental symbol. In the case of sodium, which is a *monoisotopic element* and which therefore consists of only a single isotope, the *atomic number* (the number of protons in the nucleus) is 11 and the *mass number* (11 protons + 12 neutrons) is 23. The notation for sodium is therefore $^{23}_{11}$Na. In the case of carbon two stable isotopes are found in nature with a characteristic isotopic abundance: $^{12}_6$C (the most abundant isotope normalised to 100%) and $^{13}_6$C at 1.08%. Additionally, the radioactive carbon isotope $^{14}_6$C is only found in trace amounts, that is, making up about 1 or 1.5 atoms per 10^{12} atoms of the carbon in the atmosphere. $^{14}_6$C decays with a half-life of about 5700 years and is used for determining the age of objects containg organic material (radiocarbon dating).[12]

The isotopic distribution of naturally occurring isotopes is a valuable piece of information that is considered when it comes to the examination and interpretation of the signal patterns of ions, especially for molecular ions in mass spectra. In Table 5.1 the isotopic

Table 5.1 Isotopic composition of selected elements. The abundance of the most abundant isotope is normalised to 100%. Fluorine, phosphorus and iodine are monoisotopic elements.

Element	Mass number	%	Mass number	%	Mass number	%
H	1	100	2	0.016		
C	12	100	13	1.08		
N	14	100	15	0.36		
O	16	100	17	0.04	18	0.20
F	19	100				
Si	28	100	29	5.07	30	3.31
P	31	100				
S	32	100	33	0.78	34	4.39
Cl	35	100			37	32.4
Br	79	100			81	97.5
I	127	100				

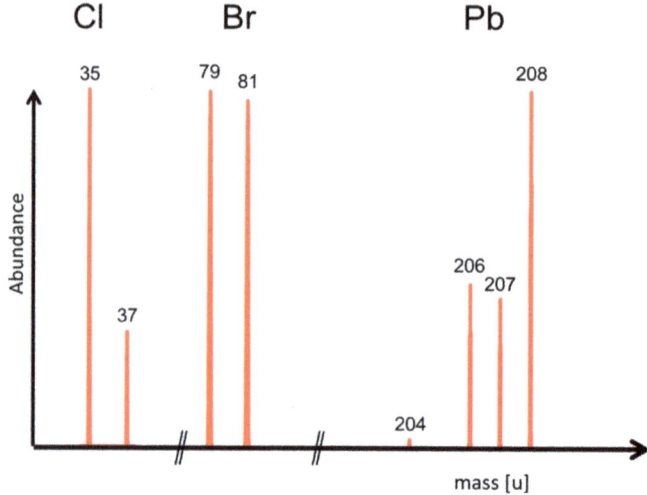

Figure 5.4 Isotopic patterns of chlorine, bromine and lead. The integer masses of the isotopes are given as the unified atomic mass unit [u] or Dalton [Da]. The *unified atomic mass unit* **u** or Dalton **Da** is defined as 1/12 of the mass of a carbon atom of the ^{12}C isotope; 1 u = 1.66 × 10^{-27} kg. The Isotopic distribution was computed using the Universal Mass Calculator (UMC) (Version 3.7.2.58) ©Matthias Letzel University Münster, Germany.

compositions of common elements are listed. However, the presentation of isotopic distributions in a bar graph format illustrates the natural abundances of the isotopes more clearly (Figure 5.4).

5.2.4 Atomic Mass and the Unified Atomic Mass Unit

5.2.4.1 *Is a MS Instrument a Sensitive Balance?*

To illustrate how small atoms are here are just two examples: 1 g of hydrogen contains 602 204 530 000 000 000 000 000 H-Atoms and 1 molecule of H_2O has a mass of 29.9 × 10^{-24} g. Hence, a *unified atomic mass unit* **u** was introduced for convenience as the mass of atoms is so tiny that to express it in kilograms would be cumbersome. The *unified atomic mass unit* **u** (which is also called a Dalton with the symbol **Da**) is defined as 1/12 of the mass of a carbon atom of the ^{12}C isotope; accordingly, 1 u = 1.66 × 10^{-27} kg. By convention the mass of an atom of the ^{12}C isotope is exactly 12 u.

5.2.5 Nominal Mass

To approximate the mass of a molecule the integer masses of the elements in the compound are summed up according to the formula. In a simple case of methane CH_4 for example, we take 12 u for ^{12}C carbon and four times the mass of ^{1}H hydrogen, which is 1 u and get 16 u as the rounded mass for a molecule of methane. In doing so, we have calculated the *nominal mass* of methane. The *nominal mass* is defined as the integer mass of the most abundant natural isotope of an element. For a molecule the *nominal mass* is the rounded mass according to the empirical formula, based on the masses of the most abundant isotopes of the elements present. It is important to note that in organic molecules containing (H, C, N, O, S, Si, and P) the lightest isotopes are the most abundant ones (*e.g.* ^{1}H for hydrogen, ^{12}C for carbon, ^{14}N for nitrogen, ^{16}O for oxygen and ^{32}S for sulphur). *This fact is not true for metals!*

In the presence of metals, we have to be careful and must remember the definition of the nominal mass, as the most abundant isotope is used to calculate the nominal mass of the molecule, ion or complex. To highlight this, we consider tetra-ethyl-lead $(C_2H_5)_4Pb$, which was used for decades as an antiknock additive for car fuels. The most abundant isotope of lead is ^{208}Pb and according to the definition that isotope is selected for the calculation of the *nominal mass* of PbC_8H_{20}: 208 u + 8 × 12 u + 20 × 1 u = 324 u. The lightest isotope of lead is (^{204}Pb) as illustrated in Figure 5.5.

5.2.6 What Is the *Mass-to-charge* Ratio?

In MS ions are formed by an ionisation process (*e.g.* EI) or (CI), see Sections 5.4.1 and 5.4.2) and these charged particles are characterised by their *mass-to-charge* ratio (*m/z*). This is the value that is exclusively measured in MS and not just the mass. The *mass-to-charge ratio* is defined as a three-character symbol. The *m/z* is used to denote the *dimensionless* quantity formed by dividing the mass of an ion by the *unified atomic mass unit* and also by its *charge number* (regardless of the sign). The symbol is written in italicized lower case letters with no spaces *m/z*.[14] Sometimes one finds the unit *m/z* = Thomson (+ or −) in the literature (*Th*), which is however, not an official recommendation.[15]

The mass *m* in the *mass-to-charge* ratio *m/z* is:

$$\frac{m_{ion}(kg)}{m_{^{12}C}(kg)/12} = \frac{m_{ion}}{1u} \tag{5.1}$$

and therefore, dimensionless as the *number of Charges z is always positive*, and hence, *m/z* is always *positive* and *dimensionless*.

To clarify this, we come back to the tetra-ethyl lead $(C_2H_5)_4Pb$ example (see Figure 5.5). In an EI-MS ion source this analyte would lose an electron of very little mass (see Table 5.2), therefore the molecular ion region in the mass spectrum (*x*-axis now *mass-to-charge* ratio *m/z*

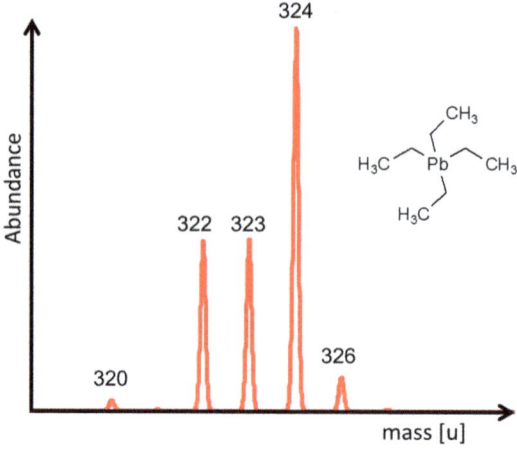

Figure 5.5 Isotopic pattern of tetra-ethyl-lead $(C_2H_5)_4Pb$. The most abundant isotope of lead is ^{208}Pb, however, the lightest isotope of lead is ^{204}Pb (see Figure 5.4). To calculate the *nominal mass* of PbC_8H_{20} (324 u) the most abundant isotopes of the elements are used. Here those are ^{208}Pb, ^{12}C and ^1H. Isotopic distribution computed using the Universal Mass Calculator (UMC) (Version 3.7.2.58) ©Matthias Letzel University Münster, Germany.

Table 5.2 Nucleons found in chemical elements with their mass and charge. The nucleus contains most of the mass of an atom as the mass of an electron is around 1/1836 that of a proton. Data obtained from ref. 13.

Nucleon	Mass [u]	Elementary charge e $e = 1.60217662 \times 10^{-19}$ coulombs
Electron (e−)	0.0005	$-1e$
Proton (p⁺)	1.007	$+1e$
Neutron (n)	1.008	0

and *y*-axis relative abundance) will reflect the composition of the ionised $(C_2H_5)_4Pb$. Consequently, we find the distribution of the individual ions of the tetra-ethyl lead molecular ion $[(C_2H_5)_4Pb]^{+\cdot}$ (a radical cation of the $(C_2H_5)_4Pb$ intact molecule with one unpaired electron, symbolized by the dot ˙, formed by the loss of just one electron) with a nominal mass at *m/z* 324.

5.2.7 Isotopic Mass and Monoisotopic Mass of a Molecule or Ion

The *isotopic mass* is the exact mass of a nuclide, which is determined relative to the mass of the ^{12}C carbon isotope, which was assigned to be the standard of the atomic mass scale with an integer value of exactly 12.00000 u.

Table 5.3 Isotopic mass of selected nuclides.

Isotope	Isotopic mass (u)
^{1}H	1.007825
^{14}N	14.003074
^{16}O	15.994915
^{19}F	18.998495
^{28}Si	27.976929
^{31}P	30.973765
^{32}S	31.972073
^{35}Cl	34.968851
^{79}Br	78.918329
^{127}I	126.904470

The isotopic masses of the nuclides shown in Table 5.3 exhibit an either positive (^{1}H) or negative deviation of the exact isotopic mass from the nominal mass, that is, a so called *mass defect*. This deviation reflects the specific stability of each individual nucleus of each atom of each chemical element having a special combination of protons and neutrons and an individual binding energy. Einstein's famous relativistic equation $E = mc^2$ (energy equals the product of mass and the speed of light squared) delivers a firm basis to describe the relationship between the mass and binding energy in any nucleus.

With the isotopic mass the *monoisotopic mass* of a molecule or an ion can be obtained. For this calculation the exact *isotopic masses* of the most abundant isotopes of the elements present in the compound of interest are taken into account. As an example, we have chosen the alkane $C_{20}H_{42}$. The *monoisotopic molecular mass* is 282.3286 u according to the composition $^{12}C_{20}{}^1H_{42}$ and the *nominal mass* is therefore 282 u. The *monoisotopic molecular mass* of tetra-ethyl-lead is calculated accordingly and leads to 324.1332 u for $^{208}Pb^{12}C_8{}^1H_{20}$ (see also Figures 5.4 and 5.5). When measured using high mass resolution capability instruments such as double focusing sector field instruments, Orbitraps[16] or Fourier transform ion cyclotron resonance FT-ICR-MS instruments,[17] accurate ion masses are measured and compared to the *monoisotopic molecular mass* calculated with the isotopic masses of the most abundant isotopes such as ^{1}H = 1.00783, ^{14}N = 14.00307 and ^{16}O = 15.9949.[6,7,18] In the case of low resolution mass spectrometers which give only integer mass accuracy, the accurate molecular masses are of academic interest only. However, when high mass resolution MS is used, the accurate masses are of importance as they can yield powerful information on the empirical formulae of parent ions and fragment ions and therefore also of the neutral losses (Δm between two ion signals).

5.2.8   Relative Atomic and Molecular Mass

5.2.8.1   *What Do I Need to Remember When Looking at Mass Spectra?*

As most of the mass within an atom is contained within the nucleus, it is essentially a measure of the number of protons and neutrons within an atom (see Table 5.2). As we have seen, the mass of an electron is significantly smaller than the mass of the protons and neutrons and their contribution to the atomic mass is usually ignored in MS. As discussed above, the *atomic number* (the number of protons in the respective nucleus) is the fundamental property of each element in the periodic table. The number of neutrons can be different and this gives rise to *isotopes. Monoisotopic elements* consist of only a single isotope (fluorine, phosphorus, iodine, sodium) but elements can have several, naturally occurring isotopes. Especially heavy metals (lead, cadmium, ruthenium, palladium *etc.*) and also the halogens chlorine and bromine (Figures 5.4 and 5.5) have characteristic isotopic patterns. The *relative atomic mass* (A_r) of a chemical element is the weighted average mass of the *isotopic masses* of its naturally occurring isotopes to an integer reference, for example, ^{12}C.

To clarify the term *relative atomic mass* of an element, chlorine can serve as a good example. There are two abundant isotopes of chlorine, ^{35}Cl and ^{37}Cl as shown in Table 5.1, which are naturally found with 75.78% and 24.22% of the total amount of Cl or having relative abundances of 100% and 31.96%, respectively (see Table 5.1). The weighted average of their isotopic masses (^{35}Cl: 34.9688 u and ^{37}Cl: 36.9658 u) delivers the *relative atomic mass* of chlorine, which is therefore 35.4528 u. This is the mass which one uses when volumetric or gravimetric calculations are carried out and that is listed in the *periodic table of elements* (see below). When looking at the chlorine molecule Cl_2 a *relative molecular mass* (m_r) of 2×35.45 u = 70.9 u is evident. The product of the *relative molecular mass* with the Avogadro constant $(N_A = 6.022 \times 10^{23})$ yields the molecular mass M_r of Cl_2 to be 70.9 [g mol^{-1}]. In other words, the molecular mass M_r [g mol^{-1}] is the mass of 1 mol atoms or molecules in g. Consequently, 12 g of the ^{12}C carbon isotope contain 1 mol or 6.022×10^{23} carbon atoms, as a single atom of $^{12}_6C$ has a mass of 12 u. This interrelation connects the microscopic world, for example, the mass of one molecule or atom with the macroscopic *amount of substance*. The international system of units (SI) defines the amount of a substance to be proportional to the number of elementary entities (atoms, molecules or ions) present. The SI unit for the amount of substance is the mole. Avogadro's constant (or number) is the number of elementary entities present in 1 mole (N_A = 6.022 × 10^{23}). A mole of a substance is the mass in g that contains 6.022 × 10^{23}particles as discussed above. 16 g of methane contains 6.022 × 10^{23} methane molecules.

It is important to note, that in the MS of small ions (nominal mass below 800 u), such as those typically found in GC-MS, the *relative molecular mass* (m_r) is not particularly helpful as we will be looking at signal patterns that reflect an individual isotopic composition of ions and hence, *the monoisotopic masses* of the ions are the important values to rely on, as we will see later on.

The elements are conveniently grouped according to their properties in the *periodic table of elements*. The periodic table is a convenient way to arrange the elements, ordered by their atomic number, electron configuration and periodic or recurring chemical properties. Each column or group contains elements with roughly similar behaviour, however, the chemical characteristics change. For example, group 1 contains the alkali metals, and group 2 the alkaline earth metals. An instructive web resource on the periodic table of chemical elements can be found here: http://www.rsc.org/periodic-table. The periodic table shows the symbol for each element, together with the atomic number and the *relative atomic mass*.

5.3 Basic Aspects of MS Instrumentation and GC-MS

5.3.1 Are All Compounds Suited for GC-MS Analysis?

As we have seen in previous chapters, compounds that are amenable to GC analysis have to be sufficiently volatile to be instantaneously vaporised in the injector before they enter the column. This prerequisite limits GC to less polar compounds, which only have weak intermolecular interactions that allow vaporisation at a low boiling point. Additionally, the molecular mass of GC amenable analytes needs to be low (typically below 800 u without using a special instrument set-up) to facilitate thermal vaporisation without decomposition *via* pyrolysis as discussed in the preceding chapters. However, it is possible to reduce the polarity of moderately polar analytes by appropriate derivatisation steps that ease the vaporisation (alcohols to silyl-ethers; primary and secondary amines to tertiary methyl amines; carboxylic acids to esters *etc.*). After separation of the injected mixtures in the gas chromatograph the gaseous fractions elute from the chromatographic column and are guided through an interface into the MS part of the

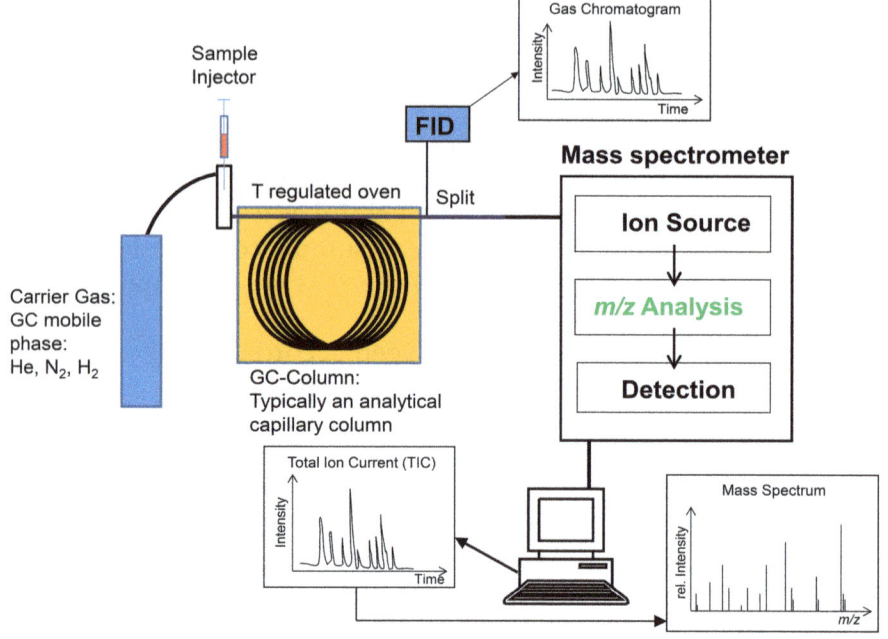

Figure 5.6   Schematic diagram of a GC-MS instrument equipped with a separate GC detector (in this case a flame ionisation detector (FID)).

instrument as illustrated in Figure 5.6. After pumping away most of the carrier gas used for GC separation the neutral compounds are ionised in the ion source component of the mass spectrometer.

5.3.2   How Do I Select the Ionisation to Fit to a GC Front End Separation?

In standard GC-MS instrumentation there are typically two options to ionise the analytes for subsequent mass spectrometric analysis: (a) *via* electron ionisation (formerly called electron impact (EI, see Section 5.4.1)); and (b) *via* chemical ionisation (CI, see Section 5.4.2), as both methods rely on volatile, low molecular mass nonpolar analytes that are ionised in the gas phase.

As illustrated in Figure 5.7, polar (typically water soluble) and macromolecular compounds are not suited for EI and CI, as analytes with these characteristics will just not reach the gas phase *via* heating. Their vapour pressure is too low even at elevated temperatures and reduced pressure conditions owing to their strong intermolecular interactions (the coulombic interactions of ions, of dipole-ion and of dipole–dipole interactions). However, alternative ionisation methods such as electrospray ionisation (ESI) are ideally suited for polar and

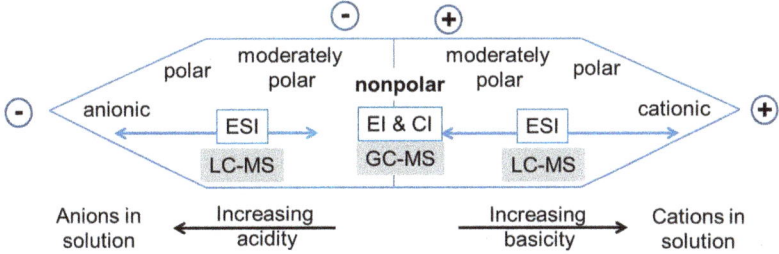

Figure 5.7   Ionisation-continuum diagram showing the regions of effective applicability for GC-MS and LC-MS, including EI, and CI as well as ESI. Adapted
from ref. 25 with permission from American Chemical Society, Copyright 2001.

macromolecular analytes and therefore also allow liquid chromatography (LC)-MS coupling. A discussion of these ionisation methods
goes beyond the focus of this short introductory book – the interested
reader is directed to the vast literature covering ESI and LC-MS.[19–22]

5.3.3   How Do I Ensure That the Mass Spectrometer Will Produce Good Spectra?

Before carrying out an analysis, you must ensure that the mass spectrometer is operating optimally in order to achieve high quality data.
Additionally, if you are working to any procedure or accreditation
standards, you must be able to prove that the instrument is working
correctly and generating good quality data in order for your results
to be acceptable. Only then can you be sure that the instrument will
produce spectra that can be library searched and will provide the correct mass and abundance information for the molecular ions, isotopes and fragments. You will also ensure that the mass spectrometer
is free from contaminants and is leak free. The way in which this is
done is to ensure that the instrument is tuned and calibrated before
carrying out an analysis. In tuning the instrument, you are ensuring
that the correct voltages are applied to the ion lenses and other electrostatic components within the ion source and the mass analyser
(see Section 5.4.1 on EI-MS). Amongst the parameters optimized are
the repeller voltage, ion focus voltage and electron multiplier voltage (see Figure 5.8). Atomic mass unit (AMU) gain or radio frequency
(RF) gain and RF offset are the quadrupole parameters affecting the
resolution and sensitivity of the mass analyser in quadrupole instruments (see Section 5.5.3 on quadrupole mass analyzers). Parameters
for instruments with different mass analysers are outside the scope
of this text, but the optimization of resolution (see Section 5.5.2.1)
and sensitivity will be the desired outcome for any instrument

type. The instrument should be first tuned in the EI mode, even if carrying out work using CI (Ionization is discussed in the following sections). The instrumental response is optimized to a standard tuning compound which is often perfluorotributylamine (PFTBA, FC-43; $C_{12}F_{27}N$). This is a compound chosen owing to its stability, high boiling point and its ability to give a range of fragment ions across the mass range required for most GC-MS analysis – up to its molecular ion $[M]^{+\bullet}$ at m/z 671. Additionally, it only has ^{19}F and ^{15}N isotopes therefore the spectra are simple and easy to interpret. The tuning compound is located in a glass vial directly connected to the vacuum of the mass analyser and is automatically allowed into the mass spectrometer *via* a valve during tuning. The ions of interest during the tuning process typically appear at m/z 69, 131, 219, 414 and 502. These ions are produced with a small mass defect and enable the mass axis to be accurately calibrated across the mass range of interest. This calibrant is also used to ensure that there is sufficient resolution between the masses. There is naturally a tradeoff between the resolution and sensitivity with narrower mass windows resulting in reduced sensitivity. This is described in Section 5.5.3 and illustrated in Figure 5.20. The same tuning compound can also be used as a calibration gas when running in the negative ion CI mode, but remember that it is best to ensure your instrument is well tuned in the EI mode prior to starting CI work. Tuning a modern mass spectrometer is quite straight forward as all modern mass spectrometers have an automated tuning program. The automatic tuning will result in a compromise between resolution and sensitivity with a good mass accuracy across the entire mass range. During the autotune routine, the ion source parameters are set to ensure that the optimum abundances and ion ratios of each tuning compound fragment are formed. The quadrupole parameters, AMU gain and offset are optimized to ensure the optimum compromise between peak width and peak intensity is achieved, together with the correct mass axis calibration. The electron multiplier voltage is also optimized. The filament current may also be optimized. During autotuning, the electron voltage always remains at 70 eV, but of course this parameter may be changed manually if so desired (see Section 5.4.1 and Figure 5.11). The operator may choose to manually tune the instrument by varying the various parameters and choosing a set of parameters which may be more suitable for a particular analytical requirement. It is often advisable to carry out an autotune on your instrument first and then change the parameters manually to optimize the instrument for your particular requirements. Mass peak widths can be optimized by adjusting the AMU gain and AMU offset values. Adjusting the AMU gain has the largest effect on the higher masses whilst the AMU offset effects the peak widths equally across the mass range. Mass calibration can be adjusted by changing the mass gain and mass offset parameters.

In some instruments, these parameters may not be visible to the operator, but the mass calibration can be affected by inputting certain values into the relevant parts of the tuning software. Once the parameters have been optimized, the settings can be saved in your own user created tune file. It is strongly advised that you do not overwrite tune files so that you can go back to using previous versions if necessary.

When tuning high resolution instruments, the fragment ions will be assigned their accurate masses. For example, the fragment ion with a composition of CF_3^+ has a nominal mass of 69 u and an accurate mass of 68.9952 u (note the negative mass defect, which originates from the unique mass of the ^{19}F isotope). The tuning function should be run at the start of each working day and for critical pieces of work, the user may wish to tune the instrument prior to each analysis. It is very useful to keep a record of the optimum set of instrumental parameters when the instrument is new and also after it has been serviced. It is advised that the user should print out or store the tune page so that the previous tune conditions can be referred to. This gives a visual record of the intensities, peak shapes, mass accuracy and resolution of the instrument when it is working optimally. In this way, you can track the performance of the instrument over time and monitor its performance. You will be able to predict when the ion source requires cleaning and even gauge when the instrument should be serviced.

5.3.4 How Do I Keep My Mass Spectrometer Running Optimally?

If you keep track of your tuning parameters you will be able to determine any reduction in the sensitivity, spectral resolution and peak shape before they have a material effect on your analysis. You will periodically need to clean your mass spectrometer ion source. The frequency of this will be dependent on the number and type of samples you analyse. To keep your instrument working optimally for as long as possible it is always good practice to minimize the amount of sample that goes into the mass spectrometer. Therefore, you may wish to use as high a split ratio for your GC injector as you can. You may also wish to employ a solvent delay, which ensures the filaments are not on until after the solvent peak has eluted. Of course, if your compound of interest elutes before the solvent, you may wish to arrange for your filaments to be on at the beginning of the run and then off whilst the solvent elutes. In any case you need to know the dead volume/ dead time of your GC procedure to adjust these parameters accordingly. Both of these scenarios can be set up in the acquisition software for your instrument. Sample clean up procedures, such as filtration, solvent extraction, solid phase extraction (SPE) or solid phase micro extraction (SPME) may also by employed which will minimize the amount of matrix which gets into the instrument. However, you will need to clean the ion source and possibly the pre-quadrupoles periodically to ensure that you maintain your instrument in good condition. You will also need to replace the filaments as they become less effective. Cleaning the ion source components and replacing the filaments is a straightforward procedure and you should have a maintenance schedule which meets the requirements of your analysis. Your mass spectrometer manual will describe the best approach to this maintenance. You will also need to ensure that the oil in the roughing pump is topped up and changed according to the instrument maintenance schedule (see your instrument manual). You will also need to carry out a major instrument service, usually on an annual basis. This is best done by an appropriately trained engineer. If maintained correctly, your instrument will give you many years of good analyses.

5.3.5   Why Do I Need a GC for the Analysis of a Complex Mixture of Compounds?

To avoid overlaid mass spectra of two or more analytes, that are vaporised and ionised at the same time, a chromatographic separation prior to MS analysis (either GC or LC) is mandatory. Composite spectra with more than one compound present will overlay and become complicated, respectively, preventing reasonable interpretation and correct structure assignment.

5.3.6   Are There Artefacts From the Analytical Technique – What Should I Be Aware of?

Owing to the harsh conditions in the GC inlet (typical temperature 150–250 °C) thermal degradation reactions can take place that actually change the chemical structure of the analytes before they even reach the column or the MS stage. These thermal reactions include isomerisation reactions such as the shift of double bonds in alicyclic ring systems and the thermal loss of small neutrals owing to the interaction with hot metal surfaces in the inlet system of the MS or the GC. Molecules such as CO_2; CO, HX (H_2O, HCl *etc.*) can be lost. As an

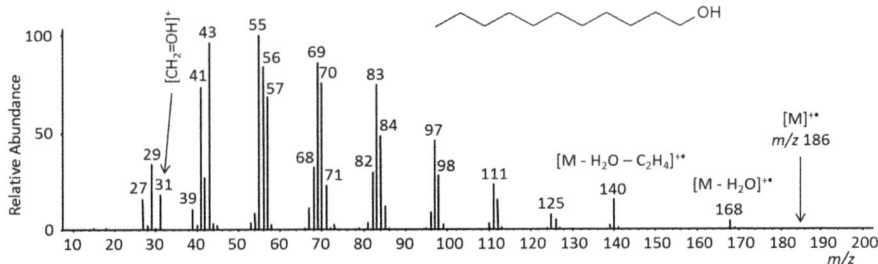

Figure 5.8 EI-MS spectrum of 1-dodecanol. Extra attention is recommended when examining the EI-spectra of alkenes as they are very similar to those of 1-alcohols owing to the facilitated loss of H₂O. Spectrum taken from the NIST Mass Spectral Library, Version 2.0f, 2008. Reproduced with permission.

example, we present the EI spectrum of 1-dodecanol (Figure 5.8) that is very similar to the respective alkenes. This is because the saturated alcohols such as the former easily lose H_2O and the resulting product of this elimination reaction delivers an EI-spectrum that is hard to distinguish from 1-dodecene. Additionally, the absence of the molecular ion signal at m/z 186 complicates the correct identification of this saturated alcohol. A recommended confirmative experiment is the derivatisation of the alcohol to an ether, for example a trimethylsilylether and comparison of the respective EI-MS of the trimethylsilylether with the original data set.[6,7,18,24,25]

5.4 How Are Ions Generated in a MS Instrument?

5.4.1 Electron (Impact) Ionisation

In EI the ions are formed at very low pressure, around 10^{-5}–10^{-6} Torr in the ion source upon bombardment with fast electrons (kinetic energy of ionising electrons in EI $E_{kin} = 70$ eV; see Figure 5.9). The low pressure effectively prevents any collisions of the ions with gaseous molecules or atoms and hence, the energy that is imparted on the molecular ions by ionisation is not moderated by any means. Consequently, the internal energy of the molecular ions comes solely from the ionisation event and is the energy reservoir for any subsequent fragmentation processes. The molecular ions can dissociate in *monomolecular* reactions that lead to the formation of fragment ions that are observed in the EI mass spectrum. These reactions take place in about a microsecond as this is the typical time the ions reside in the ion source prior to acceleration into the MS analyser (Figure 5.9).

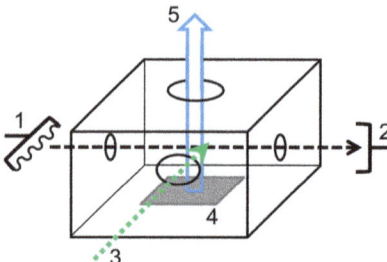

Figure 5.9   Schematic diagram of an EI ion source: (1) The filament (Re, W) emits electrons that are accelerated towards the electron trap (2) owing to a potential difference of 70 V; (3) gaseous analyte molecules enter the ion volume either from GC elution or are introduced by a direct inlet probe; (4) positively charged ion repeller electrode responsible for ion acceleration; (5) ion beam towards the analyser.

The high vacuum within the mass spectrometer is achieved by means of vacuum pumps, usually a rotary pump to pump away most of the air and a turbo molecular pump to achieve the very high vacuum necessary to avoid any scattering and deflection of the ion beam and to ensure a sufficiently long mean free path of the ions. In typical EI mass spectra the electrons, produced by a filament within the ion source, have a kinetic energy of 70 eV as they are accelerated through a potential of 70 volts $(1 \text{ eV} = 96.5 \text{ kJ mol}^{-1})$.

5.4.1.1   Why Is 70 eV Used as the Electron Energy in Routine EI MS?

The *ionisation energy* (IE) that is, the minimum energy needed to remove an electron from the highest occupied molecular orbital, of most organic molecules lies in the range 7 to 11 eV. In EI MS the fast 70 eV electrons interact with the gaseous molecules within the ion source and transfer some of their kinetic energy to the molecules (Figure 5.10a). The ionisation process (10^{-16} s) is very fast and much faster than the subsequent fragmentation reactions (Figure 5.10b).[24,25]

The probability of ionisation is dependent on the compound investigated, but for most organic molecules the highest probability is at around 50–70 eV. It can be seen that the ionisation efficiency (or the ion current) is very low at lower electron energies (below 10 eV) and rises rapidly to a maximum at around 50–70 eV as shown in Figure 5.11. The magnitude of the total ion current (overall ion intensity; TIC) and the extent of fragmentation are both dependent on the electron energy. For low electron energies that are just above the IE of the analyte, M, the TIC mainly consists of molecular ions (Figure 5.13a–c). At about 50 eV the total ion current includes the molecular ions and an

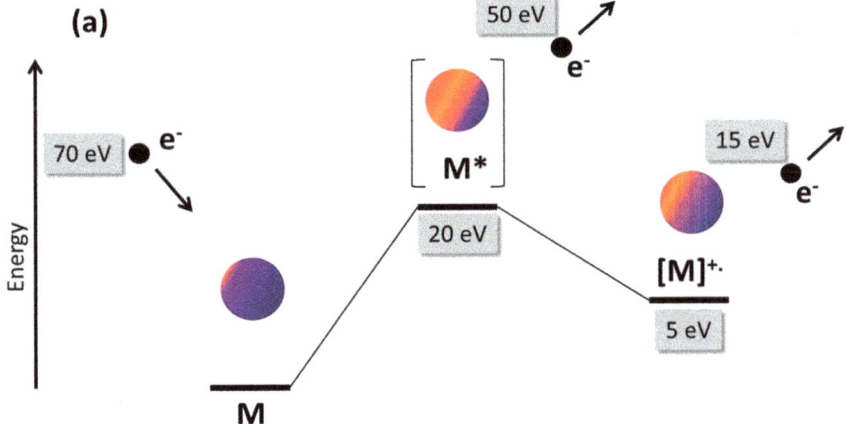

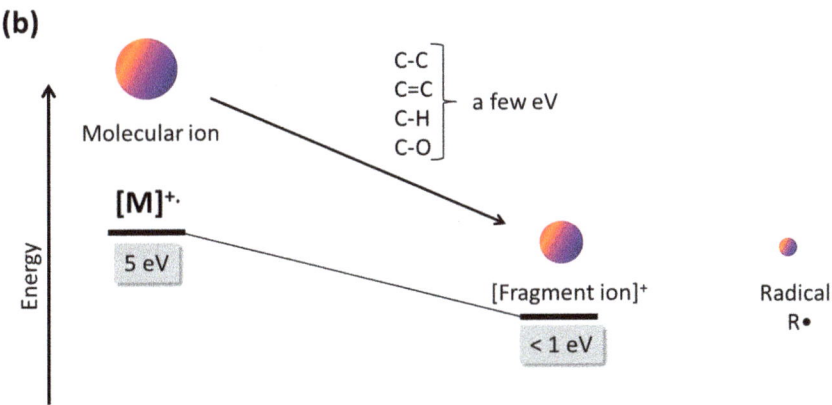

Figure 5.10 (a) Interaction of the ionising 70 eV electrons with a gaseous analyte M in an EI ion source. A small portion of the molecules become ionised (≈0.1%) and electronically and vibrationally hot molecular ions [M]⁺• with an internal energy of about 5 eV are formed (see Figure 5.11). (b) The molecular ions formed by the loss of an electron in the EI process [M]⁺• possess a substantial amount of excess energy (about 5 eV) that can lead to extensive fragmentation reactions, for example, the loss of a radical. Reproduced from ref. 24 with permission from Elsevier, Copyright 2005.

increasing contribution of fragment ions (Figure 5.11). At about 70 eV the ion current curve reaches a flat maximum that ensures reproducibility and stable experimental conditions. This is convenient when carrying out quantitative analysis at 70 eV as slight changes in the electron energy have a limited impact on the instrumental response. We note that the probability of a molecule being ionised in an electron ionisation source is very low, around 1 in 10^3 or 10^4.

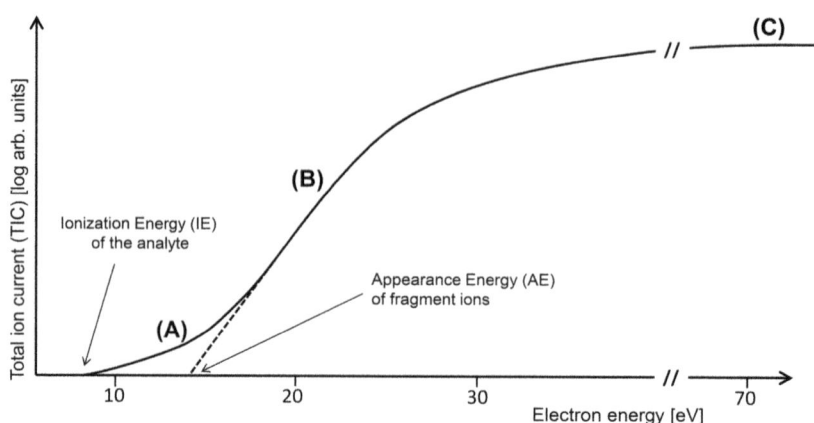

Figure 5.11   Increase of the total ion current (TIC) with increasing electron energy. (A) The threshold region after exceeding the IE of the analyte M (for the molecular ion [M]$^{+\bullet}$ the IE is roughly the appearance energy (AE)). In this electron energy region molecular ions are mainly formed. (B) Build-up region with increasing production of fragment ions owing to the increasing excess energy of the molecular ions formed. (C) Routine operation at around 70 eV electron energy with stable reproducible formation of fragment ions.[25-27]

All neutral organic molecules which are not radicals, not isotopically enriched and which do not contain any nitrogen atoms (the case of nitrogen containing molecules will be discussed separately) have an even number of electrons and an even molecular mass. It becomes obvious that when a molecular ion [M]$^{+\bullet}$ is formed by the removal of an electron, the resulting radical cation has a *mass-to-charge* ratio which is an even number. Positive ions are exclusively formed in the EI ion source and only positively charged ions are detected in the EI mass spectrum. In most cases the ions which are formed in EI sources carry a single charge ($z = 1$) therefore the m/z ratio is equal to the mass of the ion. Ions that represent the intact molecules after removal of an electron are the molecular ions [M]$^{+\bullet}$ and are open shell radical cations with an odd number of electrons (OE$^{+\bullet}$). The initial ionisation process taking place in the ion source of an EI-MS is an oxidation process.

$$M + e^- \rightarrow [M]^{+\bullet} + 2e^-$$

M is a neutral gaseous molecule, e$^-$ is an electron with a kinetic energy of 70 eV, and [M]$^{+\bullet}$ is the molecular radical cation which contains an excess of internal energy (compare with Figure 5.10a and b). In addition to the ionisation energy (IE), around 5 to 8 eV of internal energy is imparted to the molecular ion as it is formed (Figure 5.10a). The molecular ion now has an excess of energy and as there is

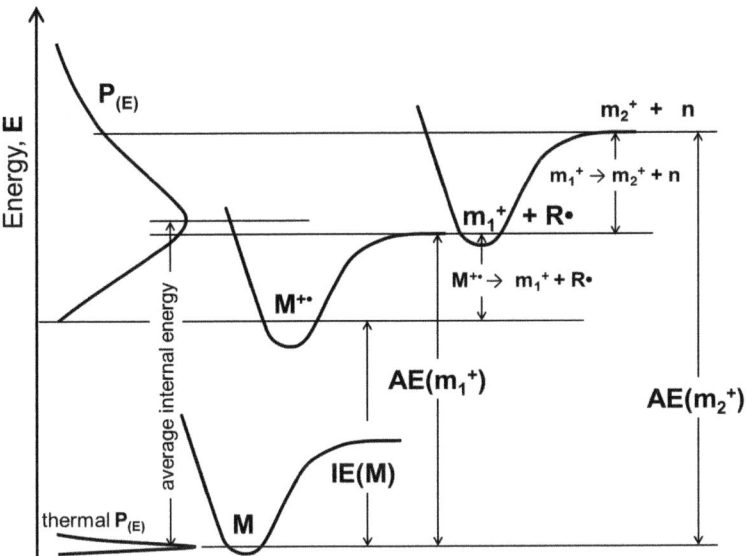

Figure 5.12 The appearance energy (AE) of a given fragment ion is the minimum energy needed to observe that ion in the mass spectrum. For a molecular ion the AE is the ionisation energy (IE). The distribution of the internal energies as a function of the energy $P_{(E)}$ of the molecular ion $M^{+\bullet}$ governs the accessible fragmentation reactions and the formation of fragment ions m_1^+ and m_2^+. The ionisation process furnishes the molecular ion with a certain amount of internal energy that can be sufficient for subsequent fragmentation (in this example formation of a fragment ion m_1^+ by loss of a radical $R^\bullet$). If this primary fragment ion is still activated enough it can further fall apart and form m_2^+ by the loss of a neutral n. However, m_2^+ could also be formed directly from a small, highly energetic portion of the molecular ions $M^{+\bullet}$ which are found on the upper end of the energy scale (above the $AE(m_2^+)$ level). All processes after ionisation, for example, isomerisation, fragmentation and so forth, take place in the ion source in a timeframe of about 10^{-6} s and are therefore kinetically controlled. Adapted from ref. 18. Reproduced with the permission of the publisher.

no possibility of collisional interactions with other species the only way the ions can lose this excess energy is *via* vibration, eventually leading to fragmentation as illustrated in Figures 5.10b and 5.12. The internal energy of molecular ions formed by EI is usually more than enough to break bonds as the typical bond strength within an organic molecule is 2–4 eV (Figure 5.10b). The molecular ion will now readily and rapidly dissociate. The extent of the fragmentation is dependent upon the stability of the molecular ion (Figure 5.12). The more stable the molecular ion, the less fragmentation will result. Conversely, less stable molecular ion species will fragment more extensively (compare with Figure 5.8).

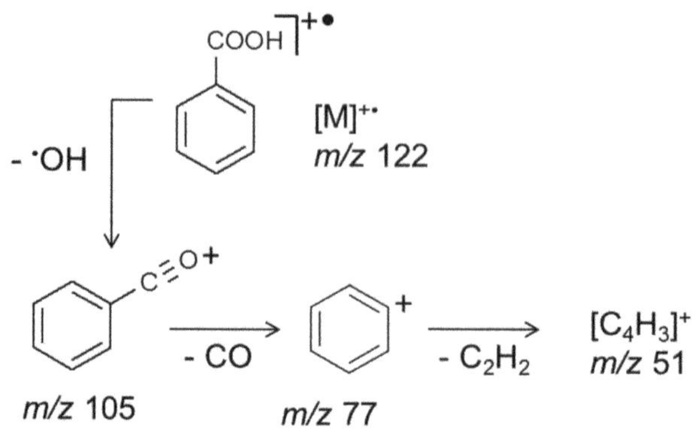

Scheme 5.1 Subsequent fragmentation pathway of the molecular ion [M]$^{+\bullet}$ of benzoic acid at *m/z* 122. The initial loss of a hydroxyl radical generates a fragment ion with an even number of electrons (EE$^+$) at *m/z* 105, that subsequently loses neutral molecules CO and C$_2$H$_2$ to ultimately form [C$_4$H$_3$]$^+$ (compare with Figure 5.13a–c).

Fragmentation occurs *via* real chemical reactions with defined mechanisms and is not a random process. The ions are present in the ion source for just a microsecond and fragmentation usually takes place within the ion source, therefore the fragmentation reactions are rapid relative to the time spent in the ion source. In Scheme 5.1 the sequential fragmentation pathway of the benzoic acid molecular ion [M]$^{+\bullet}$ at *m/z* 122 is presented. The initial loss of a hydroxyl radical generates a fragment ion with an even number of electrons (EE$^+$) at *m/z* 105, that subsequently loses neutral molecules CO and C$_2$H$_2$ to ultimately form [C$_4$H$_3$]$^+$ at *m/z* 51. This example shows that the fragmentation reactions found in EI-MS provide important structural information on the analyte. All fragment ions in the EI mass spectrum are generated from ionisation with highly energetic 70 eV electrons (Figure 5.13a). A decrease in the kinetic energy of the electrons used for ionisation reduces the excess energy imparted on the molecular ions and therefore leads to a reduced ability for fragmentation (compare Figure 5.13b and c with Figure 5.11). We will see later in many more examples that the formation of the structure characteristic fragments in 70 eV EI-MS ultimately allows structural identification and assignment. The monomolecular reactions that take place in the ion source can lead to complex EI mass spectra because many different reactions compete and can be found, as illustrated in Figure 5.14. It must be noted that a correct structural assignment on the basis of the molecular and fragment ions is only possible for a clean single component EI mass spectra. For mixture analysis a chromatographic separation is necessary and is mandatory to avoid overlaid spectra as shown in Figure 5.15.[24–26]

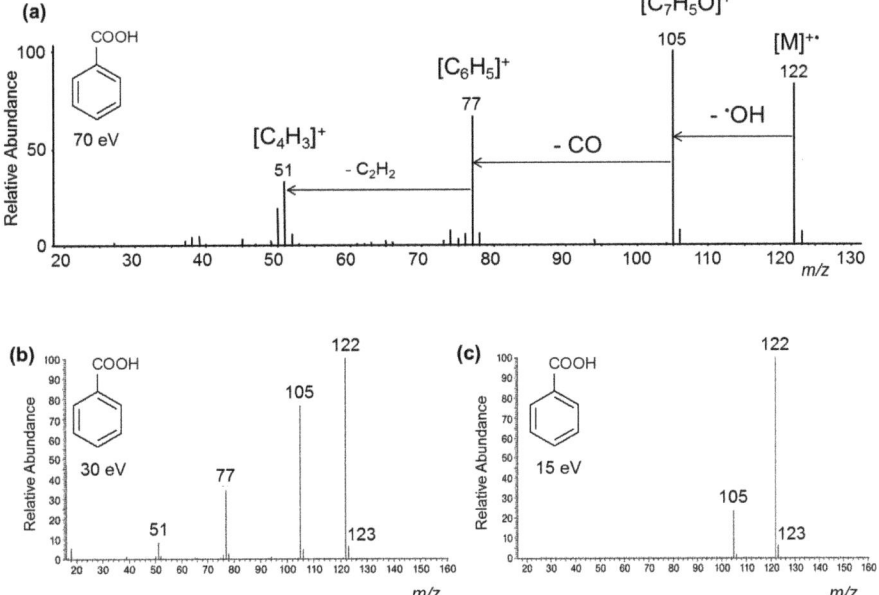

Figure 5.13   The El-mass spectra of benzoic acid acquired with different kinetic energies of the ionising electrons: (a) 70 eV (spectrum taken from the NIST Mass Spectral Library, Version 2.0f, 2008. Reproduced with permission.); (b) 30 eV; and (c) 15 eV. The formation of fragment ions is increasingly limited with the reduced kinetic energy of the ionising electrons owing to the reduced excess energy that is transferred to the molecular ion of the benzoic acid in the ionisation process. The ionisation energy of benzoic acid lies in the range of around 9.3–9.7 eV.[28] Reproduced from ref. 27 with permission from John Wiley and Sons, Copyright © 2012 Wiley-VCH Verlag GmbH & Co. KGaA.

Electron ionisation at 70 eV produces reproducible mass spectra. The spectra are effectively 'fingerprints' of the molecules and as such can be compared with EI mass spectra libraries (*e.g.* National Institute of Standards and Technology (NIST)) to enable identification of unknown analytes. Additionally, unknown spectra can be interpreted from basic knowledge of organic chemistry and MS (see Chapter 7). The disadvantage of EI-MS is that the molecules must be heated to vaporise them which can sometimes result in thermal decomposition (see Figure 5.8). In the case of molecular ions which are relatively unstable, the fragmentation can be very extensive and a molecular ion may be very small or not detectable at all (compared with the EI mass spectrum of dodecanol in Figure 5.8). To circumvent this problem the excess energy on the molecular ions can be reduced, as demonstrated in Figure 5.12a–d and in Figure 5.15. An alternative, softer ionisation

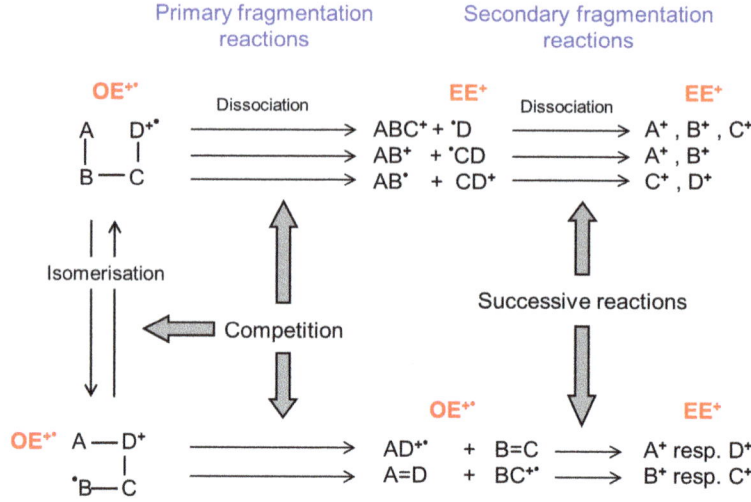

Figure 5.14 *Monomolecular* fragmentation reactions of a hypothetical linear molecular ion [A–B–C–D]⁺˙ presented schematically. The odd electron number radical cation OE⁺˙ can either rearrange and expel a neutral molecule in a rearrangement reaction (the OE⁺˙ character of the fragment ions is retained; lower pathway), or directly produce fragment ions by loss of a radical in a simple bond cleavage reaction (forming an even electron number, *i.e.*, a so called EE⁺ ion). Subsequent secondary fragmentation reactions also occur. All reactions in the EI ion source shown here are *monomolecular* and compete with each other. The energetic demand for a simple bond cleavage is typically higher than for a rearrangement reaction, but the entropic demand for the latter is higher owing to the substantially more complex transition state compared to the former.[7,18,25]

suited for low mass, nonpolar analytes is chemical ionisation, which is also widely used in GC-MS instruments and is discussed in the next section.

5.4.2 Chemical Ionisation

In CI modified EI ion sources are used that allow the introduction of an additional gas, a so-called CI gas, that is present in an large excess relative to the gaseous analyte molecules.[18,27,30] This CI gas is initially ionised by highly energetic electrons (E_{kin} 200 eV). The ion source pressures in CI are substantially higher than in EI to promote efficient *bimolecular* ion-molecule reactions between the ultimately formed plasma ions of the CI gas and the actual analyte molecules. The CI ion source (or ion volume) is tightened to adjust to elevated pressures (≈ 1 Torr) and to prevent the pumping of the CI gas. In this short overview

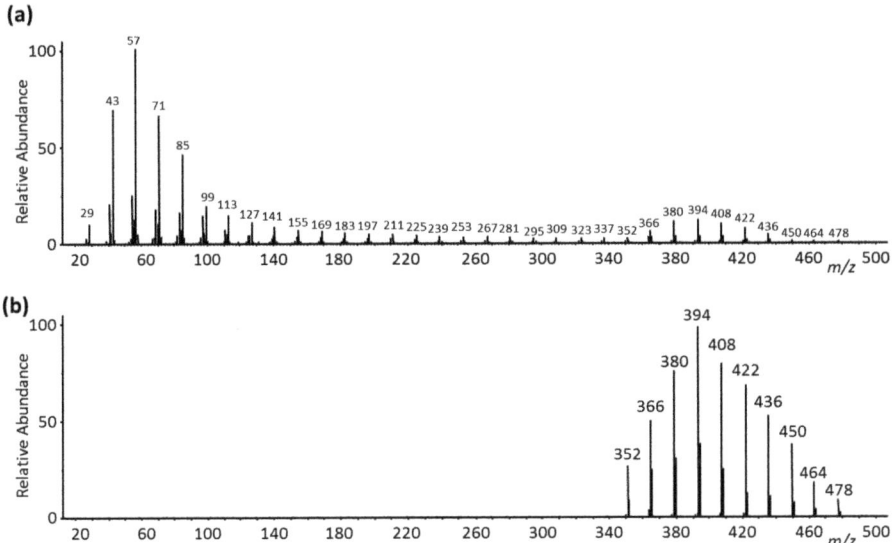

Figure 5.15 Composite EI-mass spectra of a mixture of alkane hydrocarbons at different electron energies. (a) At routine 70 eV EI electron energy many alkane fragment ions of the type $[C_nH_{2n+1}]^+$ are found with a characteristic mass difference of 14 da = CH_2 moiety (ion series with uneven ion masses: starting from *m/z* 29 to 337). (b) The minimum 17 eV EI electron energy limits the TIC to the molecular ions (compare with Figure 5.11) demonstrating the distribution of alkane compounds present in the mixture, highlighting the necessity for GC separation of analyte mixtures to allow a correct structural assignment on the basis of MS data sets. Adapted from ref. 31 with permission from John Wiley and Sons, Copyright © 1992 Wiley-VCH Verlag GmbH & Co. KGaA.

we concentrate on the formation of positive ions *via* CI (negative CI and electron capture CI can also be conducted; see references at the end of this chapter).

In CI, ionisation can be achieved in two ways depending on the selection of CI gas:

- Ionisation *via charge transfer (or charge exchange)* → odd electron number molecular ions of the analyte M are formed $[M]^{+\bullet}$;
- Ionisation *via* Brønsted acid–base chemistry: *proton transfer* → even electron number molecular ions of the analyte M are formed $[M + H]^+$.

The clear advantage of CI lies in the fact that the excess energy that is deposited on the molecular ion of either type can be controlled as necessary by appropriate selection of a CI gas. The selection is carried out on the basis of the evaluation of either the IE or the *proton affinity*

(PA). The PA of a compound is the negative reaction enthalpy of its protonation reaction:

$$PA_{(B)} = -\Delta_R H \qquad B + H^+ \rightarrow [B + H]^+$$

5.4.2.1 How Does Chemical Ionisation via Charge Transfer Work?

Ionisation *via charge transfer* can be achieved when a CI gas is selected (see Table 5.4) which delivers prominent molecular ions of the type $[B]^{+\bullet}$.[18,27,30] The ionisation takes place in the CI ion source as an ion–molecule reaction *via* a simple charge exchange:

$$M + [B]^{+\bullet} \rightarrow [M]^{+\bullet} + B$$

If the IE of the CI gas is in the range of the analyte: IE(CI-Gas) ≈ IE(Analyte), this reaction does not provide much extra energy in the molecular ions formed and this limits their fragmentation and the predominant formation of molecular ions $[M]^{+\bullet}$.

If the IE of the CI gas is substantially higher than that of the analyte: IE (CI-Gas) > IE (Analyte); then the charge transfer reaction is exothermic and excited molecular ions and fragment ions will be formed. The respective CI spectrum will not much differ from the EI-MS data.

The selection of the CI gas on the basis of an IE comparison determines the exothermicity of the ionisation reaction *via* the charge exchange, meaning that the extent of fragmentation can be influenced and limited.

Table 5.4 Reactant gases for CI *via* charge transfer. The CI gas X forms the respective [X]$^{+\bullet}$ radical cation in the CI plasma for charge exchange with the analyte molecules M. A suitable CI gas is selected on the basis of the IE value in relation to that of the analyte M.

CI gas (B)	IE [eV]
Benzene C_6H_6	9.3
Xenon, Xe	12.1
Carbon dioxide, CO_2	13.8
Carbon monoxide, CO	14.0
Nitrogen, N_2	15.3
Argon, Ar	15.8
Helium, He	24.6

5.4.2.2 How Does Chemical Ionisation via Proton Transfer Work?

Chemical Ionisation of a gaseous molecule M *via* Brønsted acid–base chemistry relies on the *proton transfer* of a CI plasma ion onto the molecule M.[18,27,30] To form the relevant plasma ions of methane, that is; CH_5^+ and $C_2H_5^+$, a number of consecutive ion–molecule reactions take place:

$$CH_4^{+\bullet} + CH_4 \rightarrow CH_5^+ + CH_3^\bullet$$

$$CH_4^{+\bullet} \rightarrow CH_3^+ + H^\bullet$$

$$CH_3^+ + CH_4 \rightarrow C_2H_5^+ + H_2$$

The relevant ion–molecule reaction between the plasma ion of the CI gas $[B + H]^+$ and the analyte M follows:

$$[B + H]^+ + M \rightarrow [M + H]^+ + B$$

Ultimately, a protonated molecular ion $[M + H]^+$ is formed which is an even electron number molecular ion (EE^+).

If the PA of the CI gas is in the range of the analyte: PA(CI-gas) $\approx$ PA(analyte), the proton exchange reaction will deliver only moderately excited molecular ions $[M + H]^+$. Consequently, the limited excitation will reduce the extent of accessible fragmentation reactions, predominantly the formation of protonated molecular ions $[M + H]^+$.

If the PA of the CI gas is substantially smaller than that of the analyte: PA(CI-gas) < PA(analyte) an exothermic proton transfer reaction is the consequence and highly excited molecular ions are generated $([M + H]^+)^*$ which will deliver fragment ions. Table 5.5 lists the proton affinities of common CI gases. It is evident that hydrogen comes with a very low $PA_{(H2)}$, as the protonated hydrogen molecule H_3^+ is very unstable and therefore very acidic. At the other end of the table, ammonia with a very high $PA_{(NH3)}$ is found, which forms the stable ammonium ion NH_4^+ which only protonates more basic analytes such as secondary or tertiary amines in CI. This can be utilized for the selective ionisation of those basic analytes in complex mixtures. Also, the number of acidic hydrogens in an analyte can be determined using ND_3 or D_2O CI on the basis of the mass shift as these protons can be exchanged:

$$R(OH)_n + NH_4^+ \rightarrow R(OH)_nH^+$$

$$R(OH)_n + ND_4^+ \rightarrow R(OD)_nD^+$$

Table 5.5 Typical CI gases used for ionisation *via* protonation. The proton affinity PA of a compound is the negative reaction enthalpy of its protonation reaction: $PA_{(B)} = -\Delta_R H$ of $B + H^+ \rightarrow [B + H]^+$.

Gas	Plasma ions	PA [kJ mol^{-1}]
H_2	H_3^+	422
CH_4	CH_5^+; $C_2H_5^+$	527
H_2O	H_3O^+	706
CH_3OH	$CH_3OH_2^+$	761
$i\text{-}C_4H_{10}$	$t\text{-}C_4H_9^+$; $C_3H_3^+$	807
NH_3	NH_4^+	840

Again, the selection of the CI gas in respect to the PA determines the exothermicity of the ionisation reaction *via* protonation, meaning the extent of fragmentation can be influenced and limited.

5.5 How Is the *Mass-to-charge (m/z)* Ratio Determined in a Mass Spectrometer?

5.5.1 Ion Acceleration

Thomson built the first mass spectrometer in the early twentieth century and was able to separate and identify isotopes of important elements such as those of the noble gas neon (^{20}Ne and ^{22}Ne; see Figure 5.2).[31] This was possible because the ions formed in the ion source region were accelerated and deflected by magnetic and electrostatic fields (Figure 5.1).

The positive potential difference U_{Acc} is the potential difference that defines the acceleration region in the ion source. The U_{Acc} is in a typical EI ion source applied to the extraction electrode = ion repeller, as shown in Figure 5.9, relative to a counter-electrode (U_{Acc} in TOF and sector instruments: kV; U_{Acc} in quadrupole instruments 25–100 V).[6,18] Positive ions are accelerated by a positive potential difference and move out of the ion volume region to the direction of the ion mass analyser. The ions pick up the respective kinetic energy E_{kin} according to the following eqn (5.2):

$$E_{kin} = zU_{Acc} = m_{ion}v^2/2 \tag{5.2}$$

In which m_{ion} is the mass of the ion and v its velocity and z the number of charges. The accelerated ions are compressed in an ion beam which is focused by electrostatic lenses and slits to allow the best resolved signal detection. The velocity v of the ions travelling through the MS instrument can be easily derived from eqn (5.3) to be:

$$v = \sqrt{\frac{2zU_{\text{Acc}}}{m_{\text{ion}}}} \quad (5.3)$$

5.5.2 Time of Flight Analysers

Time-of-flight mass spectrometers use a very simple and robust ion analyser that has been used for more than 60 years. TOF analysers rely on an evacuated flight tube and a time dependent ion detection system as Figure 5.16 illustrates.[32] Mass analysis in a TOF analyser relies on the fact that ions of different *m/z* values have the same kinetic energy, but different velocities, after acceleration out of the ion source with a constant U_{Acc} (Figure 5.16).

In TOF, ions with a high molecular mass, that is a high *m/z* ratio, will take longer to reach the detector than ions with a low *mass-to-charge* ratio *m/z* after the flight through the drift tube of a given length *L*. This provides, in principle, an unlimited mass range for TOF analyser instruments. The velocity of the ions in a TOF is $v = L\,t^{-1}$, in which *t* is the flight time from the ion source region to the detector. Using eqn (5.3) we can derive eqn (5.4) and (5.5):

$$t = \sqrt{\frac{m}{2zU_{\text{Acc}}}} * L \quad (5.4)$$

$$\frac{m}{z} = 2U_{\text{Acc}} \left(\frac{t}{L}\right)^2 \quad (5.5)$$

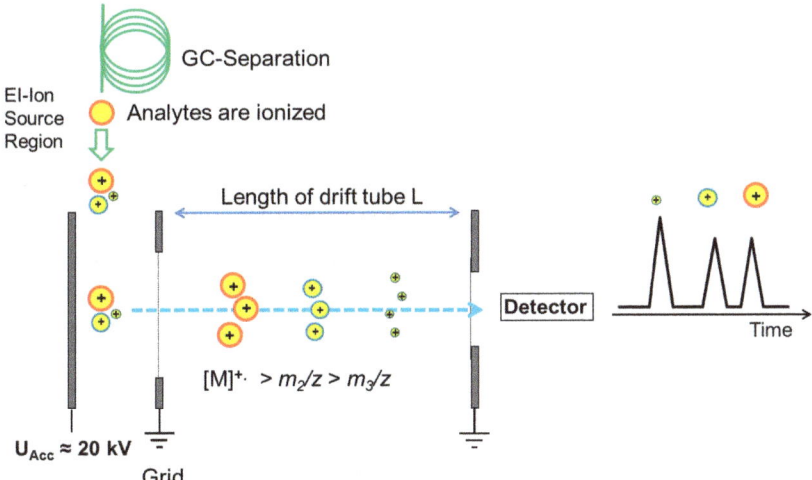

Figure 5.16 Schematic diagram of an exemplary GC-TOF instrument. The velocity of the ions is in the order of $v \approx 75$ m s^{-1} and the flight time through the TOF drift tube, $t =$ ns to µs.

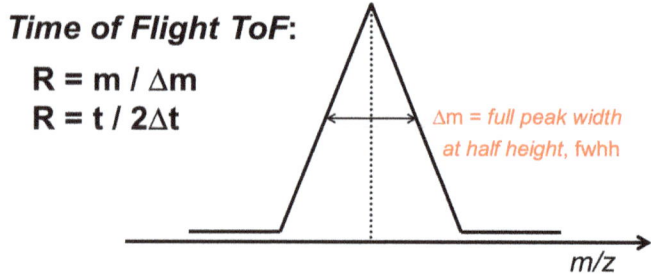

Figure 5.17 Resolution in TOFMS: the resolution is determined for every signal on the basis of the peak width at half maximum FWHM.

5.5.2.1 What Is Resolution in MS?

The general definition of resolution in MS is: $R = m/\Delta m$. From eqn (5.5) it follows that m/z is proportional to t,[2] which leads to the formula for the resolving power R in TOF analysers: $R = m/\Delta m = (1/2)\,(t/\Delta t)$. The resolution R in TOF-MS is determined for every signal on the basis of the peak width at half maximum FWHM (*full width at half maximum*) as shown in Figure 5.17.

However, the TOF analyser is a discontinuous analyser and is therefore best suited for discontinuous ion production methods such as matrix assisted laser desorption ionisation (MALDI) that can also produce macromolecular ions of polar analytes, which triggered the resurgence of TOF-MS instruments in life sciences applications (*e.g.* proteomics, genomics, lipidomics). TOF-MS instruments feature a short duty cycle, simple and robust handling and a high resolution capability (TOF analysers with an electrostatic ion reflector = reflectron; $R \geq 20\,000$ FWHM) as well as a good mass accuracy. In general, the mass resolution is inversely proportional to the acquisition rate. Those that have a high mass accuracy, such as those with $R = 60\,000$ FWHM have a slow acquisition rate, whereas those with a nominal mass can acquire at up to 500 spectra/second which makes them highly suitable for GC × GC in which the chromatographic peak widths can be as narrow as 30 ms. All this contributes to the successful development and broad use of GC-TOF-MS type instruments.[33]

5.5.3 Quadrupole Mass Analyser

The first quadrupole mass analysers were introduced by Wolfgang Paul and Hans Steinwedel around 1953.[34] Since then quadrupole MS instruments have evolved to be the most widely used benchtop GC-MS instruments because of their ease of use, small size and relatively

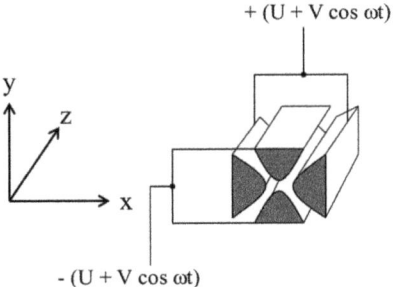

$+ (U + V \cos \omega t)$

$- (U + V \cos \omega t)$

Figure 5.18 Schematic diagram of a quadrupole set-up with hyperbolic rods. A DC voltage (*U*) and an AC voltage (*V*) with an angular frequency ω are applied to the two pairs of parallel electrodes. The resulting electrostatic time-dependent potential Φ_0 governs the ion motion in the *x/y* direction and has no influence along the *z*-axis, that is, in the direction of the original ion motion into the device.

low cost. A quadrupole mass spectrometer consists of four parallel hyperbolic or cylindrical electrodes and two opposite rods which are connected, making a set of two electrode pairs (Figure 5.18). To one pair of rods a DC voltage and a radio frequency (rf) voltage potential is applied, whereas, to the other pair of rods, a DC voltage of opposite polarity and a rf voltage with a 180° phase shift is applied. Ion separation is achieved by appropriate setting of the DC and AC voltages at the two pairs of electrodes generating a time-dependent electrostatic field which allows only an individual set of ions with a certain *m/z* ratio to travel on a stable trajectory through the quadrupole mass filter to the detector (Figure 5.19). The time-dependent potential Φ_0 applied to opposite pairs of rods is given by:

$$\pm \Phi_0 = U + V \cos \omega t \tag{5.6}$$

In which *U* is the DC voltage and $V \cos \omega t$, the time-dependent rf voltage, in which *V* is the rf amplitude and ω the angular frequency ω = $2\pi v$, with *v* being the rf (Figure 5.18).

At certain values of *U*, *V* and ω, only ions with an appropriate *m/z* ratio will travel in stable trajectories through the set of electrodes. The range of ions of different *m/z* values, capable of passing through the mass filter, depends on the ratio of *U* to *V*. All other ions will have trajectories which are unstable (*i.e.*, their respective amplitudes in the *x*- or *y*-direction exceed the limits of a stable path in the device, *e.g.* ion b in Figure 5.19) and will be lost. The equation of motion for singly-charged ions can be derived from the theory of the Mathieu equations

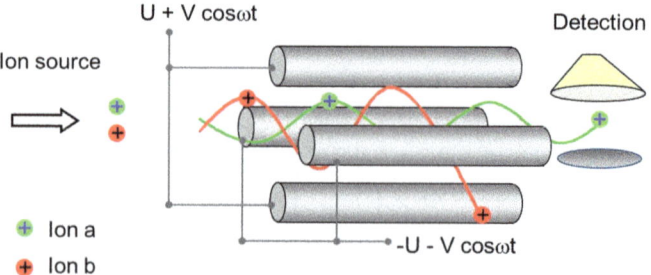

U + V cosωt

Detection

Ion source

⊕ Ion a

⊕ Ion b

-U - V cosωt

Figure 5.19 Ions with low kinetic energy in the range of 20–100 eV are introduced and follow a wiggling path through the analyser between the rods (along the z axis) under the influence of the oscillating field determined by the DC voltage (U), the AC voltage (V) and the angular frequency ω. The cation a (green) follows a stable path and reaches the detector, whereas ion b (red) is deflected and lost.

from which general parameters a_u, and q_u (u represents either x or y) and can be defined as:

$$a_u = a_x = -a_y = 8\,zU/m\omega^2 r_0^2 \tag{5.7}$$

$$q_u = q_x = -q_y = 4\,zV/m\omega^2 r_0^2 \tag{5.8}$$

in which DC voltage is U, the AC amplitude is V, m/z is the *mass-to-charge* ratio of the ion, and r_0 is half the distance between two opposite rods. Quadrupole mass analysers usually scan the mass range linearly by means of changing the values of the DC voltage (U) and the AC voltage (V) at a constant ratio $a/q = 2U/V$, while maintaining a fixed value for the angular frequency ω (Figure 5.20). The mass range of typical quadrupoles is limited which perfectly accommodates the mass range of analytes that are amenable to GC separation (molecular mass < 1200 u). Other advantages of quadrupoles are the good transmission efficiency through the mass analyser, high scan speeds, and a high sensitivity.

5.5.4 Quadrupole Ion Trap

The quadrupole ion trap (QIT) is the 3D version of the quadrupole mass filter, in which the quadrupole field affects all three x/y/z-dimensions to make ion capture, storage and ion manipulation possible. The QIT has a ring electrode and two end capping electrodes as shown in Figure 5.21, to which time-dependent potentials Φ are applied.[36] To concentrate the ions in the centre of the trap a buffer gas (usually helium $P_{He} \approx 10^{-3}$ Torr) is admitted to the trap,

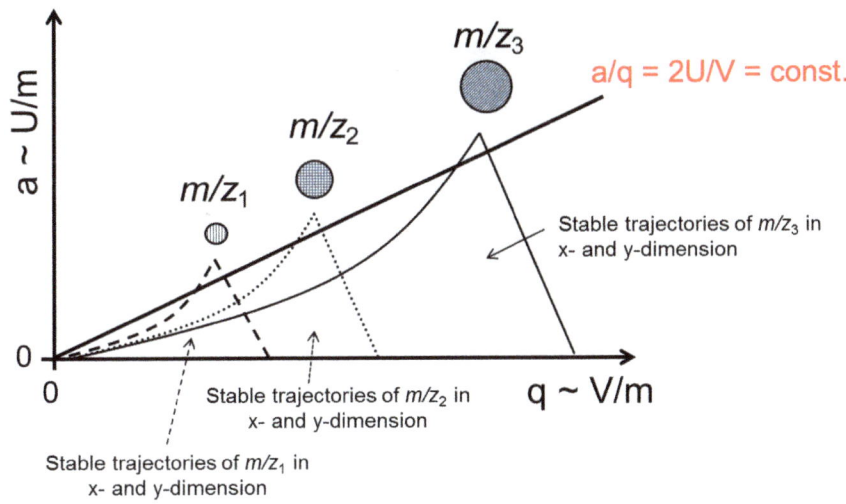

Figure 5.20   Linear scan in a quadrupole MS analyser along the $a/q = 2U/V =$ constant line (with: $m/z_1 < m/z_2 < m/z_3$). Scanning of a quadrupole analyser allows only a single ion species with a certain m/z ratio at a time to travel on a stable x/y-wiggling pathway through the device, whereas ions with inappropriate m/z ratios are deflected. An a/q-line with a steeper slope would give a higher resolution, as long as it goes through the stability areas. However, quadrupole analysers are usually only unit mass resolution devices. In the case of quadrupole operation without any contribution from the DC, for example, if $U = 0$ (only a *rf* voltage is applied) the *rf-only* quadrupole acts as a high-pass, transmitting ions above a certain cut-off mass. The *rf-only* quadrupole (q) is used for effective collision activation experiments in tandem MS, in particular in triple quadrupole instruments (QqQ).[35]

which dampens the oscillations of the ions and reduces their kinetic energy. In doing so the trapping efficiency, signal to noise ratio and the sensitivity is increased.

Ions can be stored in QIT instruments over a long period of time (up to seconds) and offer unique tandem MS capabilities, which is especially important for soft ionisation techniques such as electrospray, in which mainly molecular ions are formed. The mass range of commercial instruments reaches up to about 1000 u and these QIT devices (and also the 2D quadrupole MS bench top instruments discussed above) usually only offer unit mass resolution.[37]

5.5.5   What Is Unit Mass Resolution?

The unit resolution allows the correct separation of the signals of nominal m/z values, for example, ions at m/z 100 and at m/z 101 or m/z 500 and at m/z 501 (Figure 5.20). Space charge effects (ion–ion

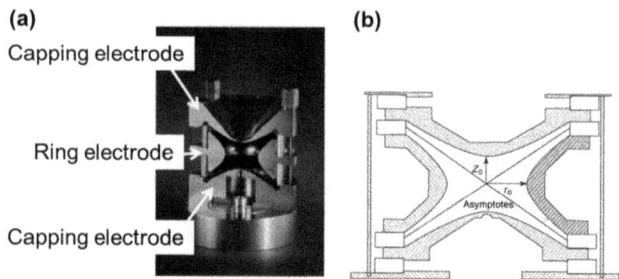

Figure 5.21   (a) Vertical cut through of a spherical quadrupole ion trap with capping electrodes on the top and bottom and the ring electrode in the centre of the ion cage device. (b) Schematic view of capping electrodes on the top and bottom and the ring electrode in the centre of the QIT. Reproduced from ref. 37 with permission from John Wiley and Sons, Copyright © 1997 John Wiley & Sons, Ltd.

coulombic interactions) reduce the accuracy of the mass assignment in QIT instruments and prevent accurate ion mass measurements (see Section 5.2.5). Even though ion–molecule reactions take place within the trap, EI spectra generally compare well to EI spectra acquired on quadrupole mass filters, but the prolonged lifetime of the QIT experiment can influence signal intensities and the competition of fragmentation pathways.

5.5.6   How Are the Separated Ions Detected?

In conventional GC-MS instruments the ions are usually detected with electron multipliers (EMs) or micro-channel plates (MCPs).

The EMs are funnel or horn-shaped and coated inside with a high secondary electron yield material. When the ion or electron hits this material it liberates electrons, the number released is proportional to the voltage, known as the detector or electron multiplier (EM) voltage. These electrons rebound and liberate yet more electrons and this is repeated throughout the length of the detector resulting in a cascade of electrons which then generates a signal.

Micro-channel plates work in a similar fashion but consist of two plates in a chevron shape with many tiny slots (micro-channels) drilled through them and coated with the material. Each slot acts as a continuous-dynode EM, with the liberated electrons cascading through the chevron where they hit a metal anode to record the total current. A detailed discussion of technical features and aspects of ion detection goes far beyond this introductory book and the interested reader is therefore directed to more specialised literature.

5.6 What Parameters Do I Need to Optimise When Using an MS?

5.6.1 How Do I Choose the Scan Range and Scan Speed for My Analysis?

When you are analysing unknown samples and would like to identify the components present in the sample, you must select a scan range that is wide enough to capture all of the spectral information which is potentially available. For example, if your sample contains a compound with a molecular mass of 200 u and your instrument is scanning from 35 to 150 u, you will not be able to identify the molecule as most of the spectral information will not be available to you. Remember that most of the important spectral information is carried by the higher mass fragments! You must therefore scan to at least 210 u to be able to see the signal pattern of the molecular ion (peak of the nominal mass and the isotopic signals) in the spectrum.

When considering the lower mass think about any background ions, for example avoid:

- Nitrogen at 28: start at 29
- Oxygen at 32: start at 33
- Argon at 40: start at 41
- Carbon dioxide at 44: start at 45

The scan speed is defined by the mass range and the cycle time. The cycle time is the time needed to complete one scan over the scan range specified. This ultimately defines how many data points per unit time you have in your chromatogram. You must allow for sufficient data points to accurately define the chromatographic peak. This is especially important when you are carrying out quantitative analysis. Too few data points will result in a poorly defined peak, indeed, you may significantly underestimate your peak area as the apex of the peak may be missed (Figure 5.22). In the most extreme cases, you may totally miss a fast eluting peak. As a general rule you need to acquire 15 to 25 data points across the width of the chromatographic peak to properly define a peak and ensure accurate quantitation. This is also important in cases in which you do not have baseline chromatographic resolution. If you do not acquire sufficient data points, the presence of partially resolved peaks may not be seen.

The acquisition rate required can be determined from the chromatogram average peak width in seconds at the baseline. An average

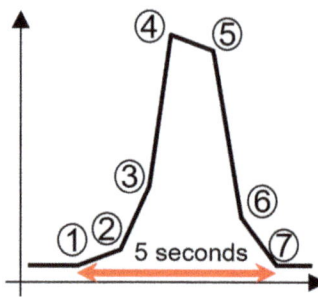

Figure 5.22 Example of a peak 5 seconds wide at the baseline with only seven data points. Courtesy of Anthias Consulting Ltd.

of 20 data points is required, therefore using the example shown in Figure 5.22:

$$\text{Acquisition rate} = \frac{20\,\text{data points or scans}}{5\,\text{seconds}} = 4\,\text{scans/second} \qquad (5.9)$$

How the acquisition rate (eqn (5.9)) is adjusted depends on the manufacturer and the mode of operation, but it is dependent on three parts:

- Mass range (full scan) or number of ions being acquired (selected ion monitoring (SIM) or multiple reaction monitoring (MRM)) for scanning instruments.
- Length of time the voltage is held to allow the ions of that *m/z* to reach the detector through: sampling rate or interval (full scan) or dwell time (SIM or MRM).
- Reset time when voltages are stabilized, this is usually not adjustable.

These can be visualised in Figure 5.23 in which the scan cycle time = 1 data point = 1 mass spectrum. The threshold value can also have an influence. This and other manufacturer-specific parameters should be fully investigated and optimised to develop the most robust and sensitive MS method.

Simultaneous acquisition instruments, such as TOF, can acquire data faster but the rate will impact the sensitivity. Therefore, the acquisition rate should be optimised as for a scanning instrument, with 15 to 25 data points across the peak.

In cases in which you know the analytes you are interested in (targeted analysis), it is pointless acquiring across a mass range which is too large. For example, if the molecular mass of your highest molecular

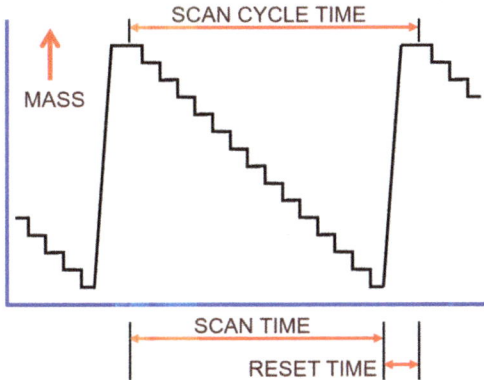

Figure 5.23 Relationship between mass range, scan time, reset time and cycle time. Courtesy of Anthias Consulting Ltd.

weight target compound is 100 u, there is little point in acquiring data up to 300 u. This will result in slower data acquisition and, in the case of the quadrupole mass analyser could reduce the overall sensitivity of your analysis.

The quadrupole mass analyser allows only one *mass-to-charge* ratio *m/z* (or a small band of *mass-to-charge* ratios) through at any given time, as the instrument scans across the mass range. Therefore, this means that if you are scanning across a wide mass range, most ions formed in the ion source will not get through the quadrupole to the detector. This represents a loss of sensitivity. For targeted quantitative analysis, it is therefore better to employ as narrow a mass range as possible. The logical extension of this reasoning is that for maximum sensitivity, we should acquire as small a number of ions as possible, only key ions present in the mass spectra of the target analytes. This is known as selected or single ion monitoring (SIM) mode. Here one or a small number of ions are collected for each analyte. If the correct ion is chosen, the signal to noise ratio (see Chapter 8) is much higher than in full scan mode. This is not only because more ions of each of the selected masses will reach the detector, but also because only the ion or ions of interest is/are being collected and none of the chemical noise or interfering ions which you would acquire in full scan mode. This can result in a reduction of two to three orders of magnitude in the detection limits. The downside of this is that you will lose a significant amount of information.

If you acquire a full spectrum, you may of course choose to plot your chromatogram using all the ions you have acquired and plot the TIC as such. Alternatively, you may choose to plot only one or a small number of ions. This may have the effect of enhancing the signal-to-noise ratio of your compound of interest, and/or enhancing the chromato-graphic resolution. Detection limits may be improved, but as you are not actually acquiring more of the ions of interest, the improvement would not be as pronounced as you would see if you were acquiring under SIM conditions.

5.6.2   How Do I Choose My Ions for SIM and Set-up My SIM Groups?

As discussed above, to obtain the highest sensitivity, a scanning instru-ment should be 'looking for' as few ions as possible at the same time, so that larger amounts of time can be spent allowing more ions of any single mass to reach the detector when acquiring data at a set acquisi-tion rate for the peak width to obtain 20 data points across the peak.

SIM mode is only used for target analysis, selecting only 2–4 ions to identify (along with the retention time) and to quantify an analyte. A unique ion (not present in coeluting peaks) is selected as the quantita-tion ion and the ratio of the remaining 1–3 ions against the quantita-tion ion are used as qualifiers. The number of ions chosen depends on the certainty required for the analysis and the presence of sufficient ions in the mass spectrum that are abundant (>10% of the base peak). Ions should be chosen that are:

- Unique to that analyte, that is not present in co-eluting mass spectra.
- Ion pairs, for example chlorine or bromine isotopes if present.
- Ions with high m/z values which are less common.
- Abundant ions, at least greater than 10% of the base peak.
- Ions representative of compound classes, for example fragment ions at m/z 76 in biphenyl spectra.

Column bleed ions (73, 207, 281, 355 and 429 are the most common for siloxane columns) and matrix ions should be avoided, sometimes it is difficult to identify interfering matrix ions until later on in the method development.

Target analytes should elute at known retention times, therefore the ions for a particular analyte only need to be acquired across that retention time window, rather than throughout the entire chromato-gram, which enables fewer ions to be scanned for at any one time. Ions for co-eluting compounds must be acquired in the same SIM group otherwise the peak front or tail may be missed. SIM groups can

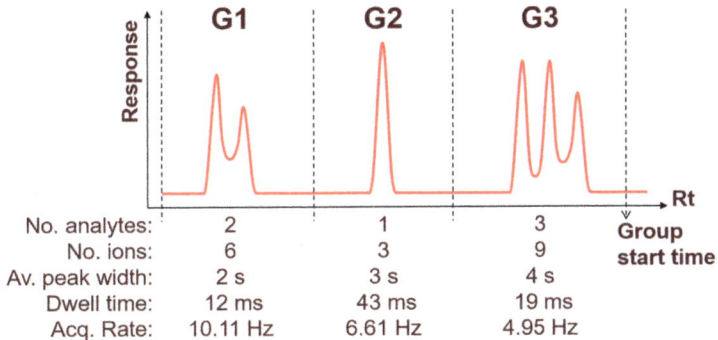

No. analytes:	2	1	3
No. ions:	6	3	9
Av. peak width:	2 s	3 s	4 s
Dwell time:	12 ms	43 ms	19 ms
Acq. Rate:	10.11 Hz	6.61 Hz	4.95 Hz

Figure 5.24 Choosing representative ions for an analyte from the mass spectrum. Courtesy of Anthias Consulting Ltd.

change where there is space on the baseline, with the objective to have as few ions as possible per group and as many groups as possible in a chromatogram – therefore good chromatographic resolution helps to achieve this!

Figure 5.24 shows an example of a chromatogram in which there are three groups of peaks, therefore three SIM groups can be created. Each group is a mini-chromatogram, with the dwell time being optimised depending on the number of analytes (hence the number of ions) and their average peak width within that retention time window. Generally, peaks broaden throughout the chromatographic run and dwell times can be optimised accurately to within 1 ms, therefore optimization of data points across a peak is more accurate than in full scan mode. It should be noted that all ions in a SIM group should have the same dwell time, and the mass of each ion should be entered with the correct number of decimal points for the instrument to achieve method sensitivity and robustness, for example modern quadrupole instruments acquire in 0.1 u increments.

5.6.3   What Temperatures Need to Be Optimised in an MS?

There are multiple temperatures throughout an MS system that need to be considered:

- *Transfer line*: this hyphenates the GC to the MS and may be controlled by the GC or the MS. It should never be higher than the column maximum isothermal temperature; nor too high to degrade analytes or for excess column bleed to be produced which will dirty the ion source. It also should not be too cool for the analytes to condense or to increase longitudinal diffusion through the column in the transfer line. As a rule of thumb, it should be 40 °C below the final oven temperature as the vacuum at the end of the transfer line effectively sucks the analytes through! Some analysts prefer to connect deactivated fused silica through the transfer line instead of the analytical column to prevent stationary phase bleed, however this also introduces an additional site for potential leaks.

- *Ion source*: the ion source temperature should be high enough to prevent the condensation of analytes, but low enough to reduce the likelihood of analyte degradation or reactions (especially when using hydrogen as a carrier gas). The temperature also depends on the ionisation technique used:
  - o EI temperatures tend to be 200–250 °C, although it has been reported that for some applications, for example PAH analysis, high ion source temperatures around 300–350 °C can improve results by reducing the adsorption effects.
  - o PCI ~ 300 °C
  - o NCI ~ 150 °C
- *Mass analyser*: depends on the type and manufacturer, but most quadrupoles are heated to 150 °C or have a pre-quad to keep the main quadrupole clean so it is not heated.

5.6.4   What Is the Solvent Delay or Cut Time and How Do I Optimise It?

The solvent in a liquid injection contains a high concentration of molecules. When they elute into the ion source the filament can become damaged or instantly break owing to the large number of molecules to be ionised, this is not a problem for most GC detectors that can cope with high concentrations. The solvent delay, also known as the solvent cut time, turns off the filament, usually along with the MS acquisition and data recording, while the solvent exits the GC column and is purged through the vacuum system. Most commonly, the solvent is the first to elute, therefore MS acquisition is not started until after the solvent delay. However, sometimes peaks elute before the solvent peak, therefore acquisition is commenced and then the filament is turned off at the solvent peak start time then back on after the solvent has eluted.

The time taken for the solvent to elute can be determined in a number of ways:

- Experimentally, by starting with a longer delay of 5 to 7 minutes, depending on the column length and flow rate, and then reducing until the solvent tail is seen.
- If the method is from another detector, that chromatogram can be used to obtain the approximate time and then optimise it as described above.
- If the MS has a vacuum gauge, the vacuum is worse while the solvent elutes owing to the high concentration, then it improves again, indicating the end of the solvent tail.

5.6.5   What Detector Parameters Require Optimisation?

The detector voltage is the last parameter to optimise, as it is much better to improve the signal by increasing the number of ions reaching the detector by optimising each step in the GC-MS method, as discussed throughout this book, than it is to purely amplify the signal. It increases the size of the peaks but can also increase the noise level if there is chemical noise from impurities, column bleed and matrix components.

Most detectors have a maximum of 2000–3000 volts and increasing the tune voltage by 50 V can double the signal. The method in which additional voltage is applied is manufacturer dependent, however applying a fixed (delta) voltage on top of the tune voltage is the most common form.

It should be noted that by using a higher detector voltage, more electrons will be knocked out to amplify the signal and this will reduce the lifetime of the detector as the electrons are used up faster. Therefore, using a voltage high enough to comfortably see the smallest peak, while minimizing the voltage to prolong the detector life is preferred.

5.6.6   What GC Parameter Considerations Are There When Hyphenating with an MS?

The vacuum pumping capacity of the MS must be known in order to decide which GC columns and flow rates can be used:

- Diffusion pumps can handle less than 2 mL min^{-1} column flows, the maximum flow rate for good sensitivity is 1 mL min^{-1}.
- Turbo molecular pumps (turbos) vary: standard turbos are able to handle 2 mL min^{-1}, but using 1–1.5 mL min^{-1} is better for optimum sensitivity, whereas high performance turbos can handle 4 mL min^{-1} but using 2 mL min^{-1} maximum greatly enhances the sensitivity. Some MS systems can cope with up to 10 mL min^{-1}.

GC columns installed into an MS should have a maximum internal diameter of 0.25 mm. This enables the use of lower flow rates to give an improved vacuum resulting in higher MS sensitivity. Two columns should only be installed into an MS with high performance turbos, the combined flow rates should be considered when optimising flow rates and it is preferable to reduce (but never turn-off) the flow rate on the column not in use to a minimum of 0.5 mL min^{-1}.

When using a scanning MS such as a quadrupole, very narrow columns, for example 0.1 mm internal diameter cannot be used. Narrow bore columns produce lovely sharp peaks, however scanning MS instruments cannot acquire data fast enough to obtain sufficient data points across them. Therefore, for GC-MS a 0.25 to 0.18 mm internal diameter GC column should be used.

Further Reading

1. J. J. Thomson, *Recollections and Reflections*, G. Bell and Sons Ltd., London, 1936.
2. E. W. Aston, *Mass Spectra and Isotopes*, Edward Arnold and Co., London, 1933.
3. J. L. Putman, *Isotopes*, Penguin Books, London, 1960.
4. A. G. Harrison, *Chemical Ionization Mass Spectrometry*, CRC Press, Boca Raton, 2nd edn, 1992.
5. J. H. Gross, *Mass Spectrometry – A Textbook*, Springer, Berlin, 2004.
6. H.-J. Hübschmann, *Handbook of GC/MS: Fundamentals and Applications*, Wiley VCH, Weinheim, 2nd edn, 2008.
7. F. W. McLafferty, F. Turecek, *Interpretation of Mass Spectra*, University Science Books, Mill Valley, 1993.
8. E. De Hoffmann, V. Stroobant, *Mass Spectrometry, Principles and Applications*, Wiley, Chichester, 2nd edn, 2002.
9. F. W. McLafferty and F. Turecek, *Interpretation of Mass Spectra*, University Science Books, Mill Valley, 1993.
10. C. Dass, *Principles and Practice of Biological Mass Spectrometry*, Wiley-Interscience, New York, 2001.

11. C. Dass, *Fundamentals of Contemporary Mass Spectrometry*, John Wiley & Sons, Chichester, 2007.
12. J. R. Chapman, *Practical Organic Mass Spectrometry*, John Wiley & Sons, Chichester, 2nd edn, 1995.
13. H.-J. Hübschmann, *Handbook of GC/MS: Fundamentals and Applications*, Wiley VCH, Weinheim, 2nd edn, 2008.
14. R. E. March, Ion Traps, in *The Encyclopedia of Mass Spectrometry*, ed. M. E. Gross and R. Caprioli, Elsevier, Amsterdam, 2004, vol. 1, pp. 144–158.

References

1. J. J. Thomson, *Philos. Mag.*, 1897, **44**, 293.
2. G. Squires, *J. Chem. Soc., Dalton Trans.*, 1998, 3893.
3. G. Audi, *Int. J. Mass Spectrom.*, 2006, **251**, 85.
4. G. Münzenberg, *Int. J. Mass Spectrom.*, 2013, **349–350**, 9.
5. J. J. Thomson, F. W. Aston, F. Soddy, T. R. Merton and F. A. Lindemann, *Proc. R. Soc. London, Ser. A*, 1921, **99**, 87.
6. E. De Hoffmann and V. Stroobant, *Mass Spectrometry, Principles and Applications*, Wiley, Chichester, 2nd edn, 2002.
7. F. W. McLafferty and F. Turecek, *Interpretation of Mass Spectra*, University Science Books, Mill Valley, 1993.
8. T. Cole, *Democritus and the Sources of Greek Anthropology*, Press of Western Reserve University, Chapel Hill, NC 1967.
9. J. Dalton, *New System of Chemical Philosophy*, London, Pt. I, 1808, Pt. II, 1810, vol. II, Pt. I, 1827, 2nd edn, Pt. I, 1842.
10. IUPAC, Compendium of Chemical Terminology, ed. A. D. McNaught and A. Wilkinson, Blackwell Science, Oxford, 2nd edn, 1997.
11. E. Rutherford, *Philos. Mag.*, 1911, **21**, 669.
12. J. R. Arnold and W. F. Libby, *Science*, 1949, **110**, 678.
13. National Institute of Standards and Technology, Proton-electron mass ratio, 214 CODATA.
14. IUPAC. Available at https://www.qmul.ac.uk/sbcs/iupac/iupac.html.
15. R. G. Cooks and A. L. Rockwood, *Rapid Commun. Mass Spectrom.*, 1991, **5**, 93.
16. Q. Hu, R. J. Noll, H. Li, A. Makarov, M. Hardman and R. G. Cooks, *J. Mass Spectrom.*, 2005, **40**, 430.
17. A. G. Marshall and T. Chen, *Int. J. Mass Spectrom.*, 2015, **377**, 410.
18. J. H. Gross, *Mass Spectrometry – A Textbook*, Springer, Berlin, 2004.
19. R. E. Ardrey, *Liquid Chromatography-Mass Spectrometry*, Wiley-VCH, Weinheim, 1993.
20. *Applied Electrospray Mass Spectrometry*, ed. B. N. Pramanik, A. K. Ganguly and M. L. Gross, Marcel Dekker, New York, 2002.
21. *Comprehensive Chromatography in Combination with Mass Spectrometry*, ed. L. Mondello, John Wiley & Sons, Hoboken, 2011.
22. *Mass Spectrometry of Proteins and Peptides*, ed. J. R. Chapman, Humana Press, Totowa, 2000.
24. E. Eljarrat and D. Barcelo, Electron Impact and Chemical Ionization, in *Encyclopedia of Analytical Science*, Elsevier, Amsterdam, 2nd edn, 2005, pp. 359–366.
25. H. Budzikiewicz and M. Schäfer, *Massenspektrometrie eine Einführung*, Wiley-VCH, Weinheim, 6th edn, 2012.

26. A. Frigerio, *Essential Aspects of Mass Spectrometry*, Halsted Press, New York, 1974.
27. H. J. Hübschmann, *Handbook of GC/MS: Fundamentals and Applications*, Wiley VCH, Weinheim, 2nd edn, 2008.
28. http://webbook.nist.gov.
30. A. G. Harrison, *Chemical Ionization Mass Spectrometry*, CRC Press, Boca Raton, 2nd edn, 1992.
31. J. J. Thomson, *Rays of Positive Electricity and their Application to Chemical Analysis*, Longmans, Green and Co., London, 1913.
32. W. C. Wiley and I. H. McLaren, *Rev. Sci. Instrum.*, 1955, **26**, 1150.
33. L. N. Williamson and M. G. Bartlett, *Biomed. Chromatogr.*, 2007, **21**, 664.
34. W. Paul and H. Steinwedel, *Z. Naturforsch.*, 1953, **8a**, 448.
35. K. Blaum, *Phys. Rep.*, 2006, **425**, 1.
36. E. Fischer, *Z. Phys.*, 1959, **156**, 1.
37. R. E. March, *J. Mass Spectrom.*, 1997, **32**, 351.

6   What Is Qualitative Analysis and How Do I Perform It?

Qualitative data analysis is the "what" is present in my sample, whilst quantitative analysis is the "how much". In qualitative analysis GC is used to confirm the presence or absence of analytes in a sample whose identities are known. GC may not lead to the positive identification of an analyte, but provides evidence of the absence of a species or if it is present below the detection limit of the method. Failure of a sample to produce a peak at the same retention time as the standard obtained under identical conditions is strong evidence of absence. GC-MS provides additional information that can be used in this process for confirmation with the mass spectrum of the analyte.

The identification of unknown analytes in a sample needs particular detectors such as mass-selective detection (MSD), ultraviolet (UV) and infrared (IR). Other data mining operations are also performed under the 'qualitative analysis' header, for example system suitability calculations to check that the instrument is performing as it should be and integration which is needed for quantitative analysis.

As discussed in Chapter 5, GC-MS data is three dimensional and the mass spectra obtained with certain MS instruments can be compared to libraries of spectra to identify analytes or can be interpreted to determine the structure of the analyte. Other types of MS such as accurate mass high resolution mass spectrometry (HRMS), enable accurate determination of the molecular ion mass which can be used to determine the empirical formula for confirmation.

Gas Chromatography-Mass Spectrometry: How Do I Get the Best Results?
By Diane Turner, Mathias Schäfer, Steven Lancaster, Imran Janmohamed, Anthony Gachanja and Jason Creasey
Published by the Royal Society of Chemistry, www.rsc.org

6.1   How Can I Use My GC-MS Chromatogram and Spectral Data?

From your GC-MS analysis, you will have two types of information:

- A chromatogram;
- Mass spectra.

A chromatogram will give a large amount of information, from enabling you to determine the health of your GC-MS instrument to being able to quantify how much of each substance is present.

A chromatogram, as shown in Figure 6.1, is a plot of the detector response against the retention time. The data is saved from the time the sample is injected and usually stops at the end of the GC run. From this, how long each sample component has spent in the GC, from injection into the inlet until being eluted from the GC column into the mass spectrometer and detected, is given by the retention time (t_r), which is taken from the apex of the Gaussian peak.

Various peak measurements can be made to determine the quality of the separation, for example resolution and peak shape (see Chapter 3). The health of the GC system can be observed from the peak shape, signal to noise ratios and the baseline.

The peak height or the peak area can be measured for quantification purposes, this will be discussed in Chapter 8.

GC-MS data is three-dimensional, within each chromatographic peak is a series of data points (see Figure 6.2) each of these data points is the ion current obtained *via* the mass spectra which are acquired several times per second.

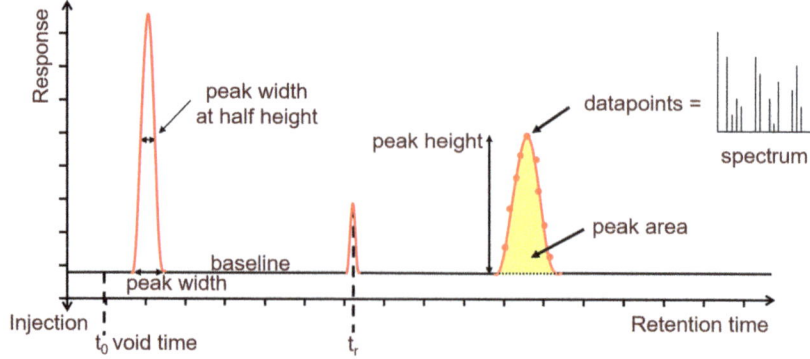

Figure 6.1   A schematic diagram of a typical GC-MS chromatogram which provides several types of information for qualitative and quantitative analysis, as well as system suitability checks. Courtesy of Anthias Consulting Ltd.

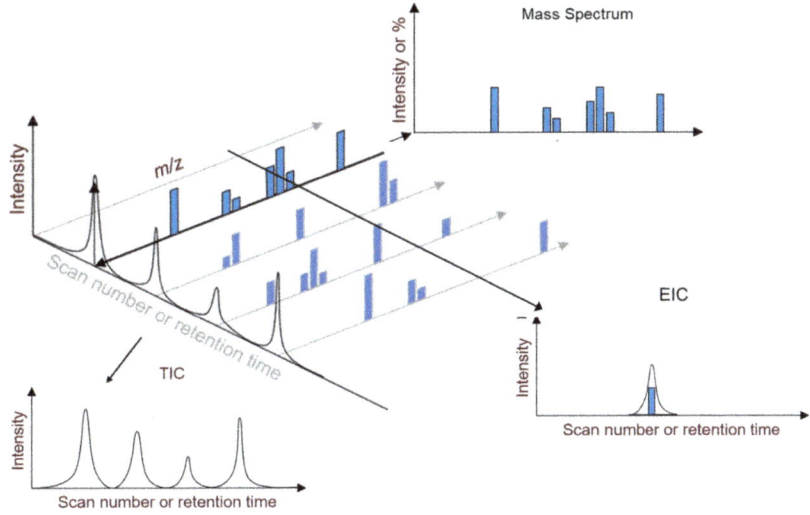

Figure 6.2 A review of GC-MS data showing the TIC, EIC and the mass spectrum. Courtesy of Anthias Consulting Ltd.

Most software quantifies compounds based on the peak area or height using the total ion count (TIC), base peak chromatogram (BPC) or from the extracted ion chromatogram (EIC) by selecting the *m/z* value(s) typical for the given compound. The mass spectrum, has previously been discussed in Chapter 5. However, Figure 6.2 illustrates the total ion chromatogram (TIC), the mass spectrum and extracted ion chromatograms that can be obtained from the 3D GC-MS data.

6.2 What Information Can I Extract from the Chromatogram About Data Quality?

Most data analysis (DA) systems will calculate and produce a report of peak measurements which can indicate the quality of the method and the health of the instrument, it may be termed as a system suitability report or enable the calculations to be made individually. However, some may not, and these can be manually calculated from the information below.

6.2.1 What Is the Baseline?

The baseline is the absence of any compounds and can provide information on the status of the instrument. It is comprised of electrical noise, impurities in the carrier gas, column bleed and any contamination from instrument parts and consumables in the carrier gas and sample flow paths.

With a mass spectrometer as the detector, a mass spectrum of the baseline at different points can be obtained to determine the source of the contamination, for example column bleed ions, solvent ions or something else. Please refer to Chapter 10 on Troubleshooting for information on individual ions.

When constructing a baseline, this can be above, through the middle or below the noise (Figure 6.3). Some DA systems give a choice, once determined this should always be used for processing data files that are related, for example when comparing two data files.

6.2.2 Chromatographic Resolution

Resolution is the measurement that is used to quantify peak separation in the chromatogram (as discussed in Chapter 3). A value greater than 1.5 indicates that the two adjacent peaks are baseline resolved. Identify a critical pair of peaks (the two most difficult to resolve) and track their resolution with each batch of samples to decide when the GC column needs to be replaced, as they are no longer adequately resolved.

6.2.3 Column Efficiency

Column efficiency is a measure of band broadening and is measured by the number of theoretical plates in a column, the higher the number the better, or by stating the plate height equivalent to a theoretical plate (HETP), the smaller the better. These terms and calculations are discussed in detail in Chapter 3.

6.2.4 Signal-to-Noise Ratio

The signal-to-noise ratio (SNR) is an important measure of the sensitivity of a method or system. Improving sensitivity has two parts – enhancing the signal and reducing noise! A simple comparison of the noise level to the peak signal level is shown in Figure 6.3.

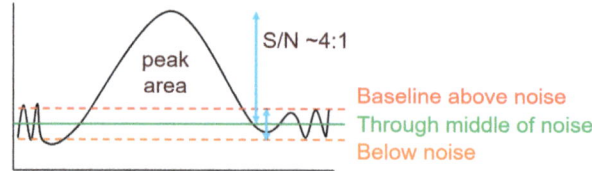

Figure 6.3 Chromatogram showing simple SNR calculation and baseline construction. Courtesy of Anthias Consulting Ltd.

Most data systems will enable the calculation of the SNR. The specifics of how to calculate it are software dependent. First, prepare a standard solution at a very low concentration, so that the peak can only just be seen with the method, if you are not sure what concentration to use, make up a series of standards at lower and lower concentrations. Load the data file in which the peak can only just be seen. Decide if the SNR calculation should be for the peak TIC or preferably the unique mass or quantitation ion and extract this. A section of the baseline close to the peak to be calculated is selected for the noise determination – it is important to ensure the area selected is clear of all peaks! Next determine the peak height from the midpoint of the noise to the peak apex. The SNR is obtained by dividing the signal by the noise, which is a unitless quantity.

In some industries, a SNR of 3 is considered to be the limit of detection (LOD) for a method and the limit of quantitation (LOQ) uses a SNR of 10. This should be decided upon for the method(s) being developed. SNR is a key parameter for trace analysis and therefore should be monitored to ensure the system remains able to detect peaks close to the LOD or LOQ.

Some DA systems offer different calculations of the SNR, for example root-mean-squared (RMS) which is defined as the square root of the mean of variances of the noise region. Therefore, investigate how your DA system calculates the value.

6.2.5   Peak Symmetry

Peak symmetry refers to how Gaussian or non-Gaussian a peak is in the chromatogram. A system set-up and optimised for an application should give symmetrical and Gaussian peaks because the processes leading to the chromatographic separation are statistical, indicating a good system. Any non-statistical processes occurring can lead to asymmetrical and non-Gaussian peaks. Fronting peaks are caused by column overloading and are described as shark fins. Tailing peaks are caused by dead-volumes, interactions, and so forth. Chapter 10 provides more detail on the causes of non-Gaussian peaks.

Asymmetry is commonly calculated as the ratio of the peak width at 10% of the peak height of the back of the peak to the front of the peak, see Figure 6.4. However, the DA system manual should be referred to for specific calculations.

The peak tailing value is calculated slightly differently from the asymmetry factor, see Figure 6.4. It is the peak width at 5% of the peak height divided by $2x$ the front width at 5% of the peak height.

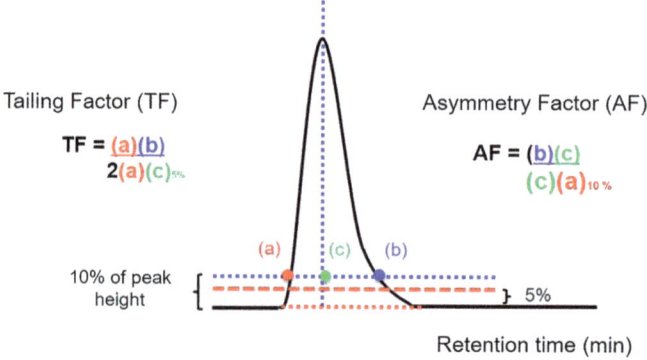

Figure 6.4 Calculation of the peak asymmetry factor and tailing factor.

The peak tailing is very useful for determining when instrument maintenance, for example inlet liner or column changes, need to be performed as it is mostly caused by activity in the system.

6.3 How Do I Determine If My Analytes Are Present or Not?

When using a GC detector, analytes can only be determined if they are present or not by their retention time and the fact that they respond to that particular detector. When using GC-MS there is more certainty in the presence (or absence) of the peak through using the retention plus the mass spectrum or representative ions. For isomers, the mass spectrum alone does not give sufficient information for confirmation, the retention time is also needed. The same can be said if only one or two ions are acquired in selected ion monitoring (SIM) mode.

The first stage of any data analysis – whether it is qualitative or quantitative, is to integrate the peaks. From there, retention time, peak areas, and so forth can be determined.

6.3.1 How Do I Integrate Peaks?

Peaks are usually found by integration to enable creation of a retention time *via* an integration algorithm. Some DA systems have multiple integration algorithms which will integrate your data more or less accurately, sometimes several need to be tried to find the one that is most suitable for the data. An integrator carries out the following basic integration process:

- Reviews the datapoints and defines the initial baseline.
- Continuously tracks the baseline.

- Observes and records any changes in slope and curvature that indicate the start of a peak.
- Follows the curve until the apex of the peak is located, creates a parabolic fit for the peak top and stores the retention time for the peak.
- Follows the curves and identifies the end time of the peak.
- Constructs a baseline between the beginning and end of the peak.
- Calculates the area, height and peak width for each peak, along with any other system suitability calculations.

6.3.2 What Affects the Peak Integration?

There are several parameters that determine which peaks are detected and reported in the software. However, the exact types and terminology are once again manufacturer software specific and even algorithm-specific, the user needs to be able to decipher these with the user manual. Some of the parameters used are:

- *Slope sensitivity*: this value can be used in peak detection sensitivity by establishing limits of the slope tangent at each time point along the axis at the peak start or end point.
- *Peak width*: this value can be used to determine the peak width of the narrowest peak and hence any peak below the value will be ignored.
- *Baseline drift*: this value can be used to observe any changes in the baseline which can be used to better integrate peaks in the column bleed profile towards the end of the run.
- *Peak number*: this value is a filter to allow the user to suggest a maximum number of peaks to be reported, usually only the largest peaks will then be tabulated.
- *Peak area or height reject*: these values are also filters allowing the user to confirm the minimum height or area of detected peaks.

Other parameters include:

- *Defining area of chromatogram to process*: start/stop integration, ignore solvent peak.
- *Defining baseline*: corrections to allow for negative peaks and column bleed.
- *Defining the peak itself*: peak width, area reject, height reject, SNR, shoulders and threshold.

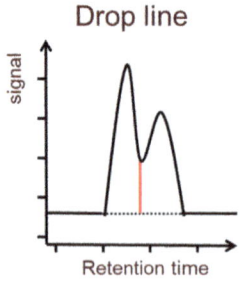
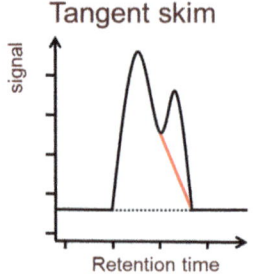
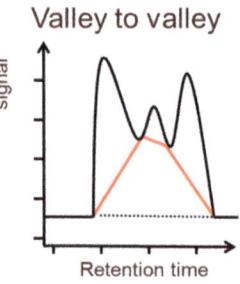

Figure 6.5   *Different ways of integrating coeluting peaks. Courtesy of Anthias Consulting Ltd.*

Whichever algorithm is used, the software will always give more accurate results with little or no manual intervention with better data – poor data gives poor results, particularly if using automatic DA methods. Hence, it is critical that the user attempts to collect the highest quality of chromatographic data for the software to integrate.

6.3.3   How Do I Integrate Coeluting Peaks?

There are several methods to integrate coeluting peaks and it is important to understand how the DA software being used will undertake this, as can be observed in Figure 6.5.

A drop line or perpendicular drop is a straight line from the valley between the two peaks to the baseline and is the most common approach that is taken for two peaks of similar sizes. Where there is a small peak on the side of a much larger peak, for example on a solvent tail, the tangent skim is a more accurate approach. Valley-to-valley is useful for peaks sitting on unresolved complex material (UCM) lumps in the baseline, such as in oil analysis, or on the column bleed baseline rise. At times, other timed events can be entered if necessary, such as negative peaks and integration start/stop, but it must be remembered that if retention times shift then these timed events must be updated.

With data from mass spectrometers, the TIC is less frequently integrated as the quantitation ion or unique mass is used. As far as possible, these should be selected so that the ion is not present in any coeluting peaks and therefore integration from baseline to baseline is much easier and more accurate.

6.3.4   Retention Time

The identification of target peaks is usually characterised by retention time. The DA system integrates all peaks within a retention time window (not the entire chromatogram) around the peak's expected retention time and then the peak closest to the expected retention time is assigned. In the case of GC-MS, the mass spectrum or the quantifier/qualifier ion ratios are also used to determine if it is a match.

A standard is required to determine the retention time of the analyte, but also the mass spectrum if the analyte is not present in the library, as shown in Figure 6.6. It is important that exactly the same method is used when acquiring the data for both the standard and the sample.

During method development, acceptable retention time shifts need to be determined, along with the correct ratios between any quantitation and qualifier ions when using SIM. The entire peak must be within the retention time window to be automatically identified, therefore any peak tailing over time must also be considered.

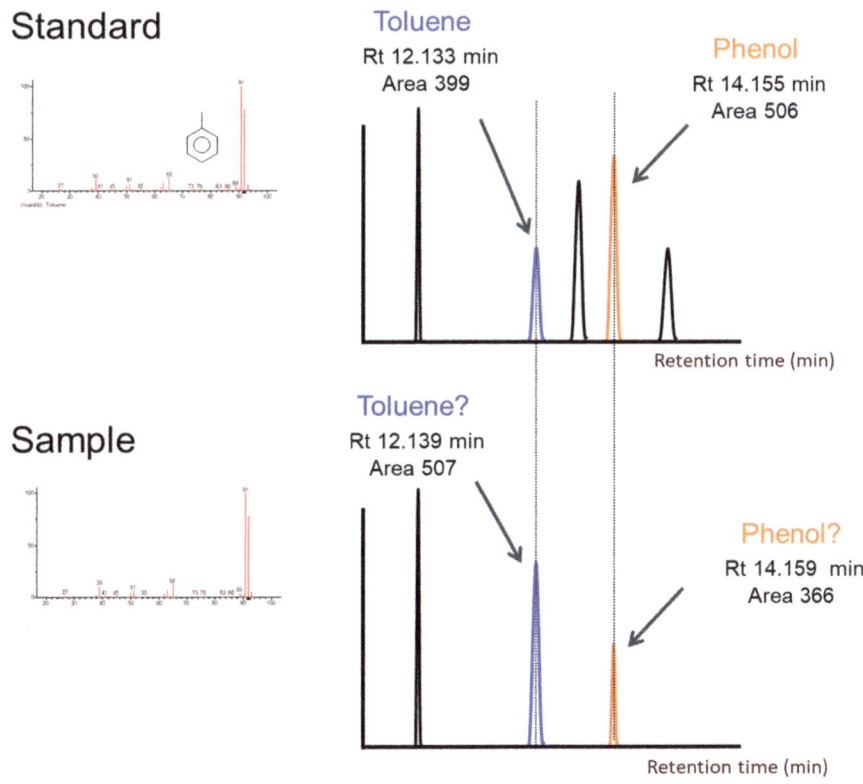

Figure 6.6 A schematic diagram of chromatograms showing how retention time and mass spectrum from a standard are used to confirm compound identification in a sample. Courtesy of Anthias Consulting Ltd.

6.3.5 Kovats Retention Index

As the spectra of isomers are often virtually the same, retention time is needed to differentiate between them. It is not always possible to obtain standards for each, but retention index data can be used if the retention indices (RI) are known. RI can also be used to compensate for retention time shifts, when a column ages or is replaced. The Kovats retention index is used to convert retention times into system-independent constants, by normalising the retention time of a compound to the retention time of an adjacent *n*-alkane eluting just before it. RI are independent of constant carrier gas pressure and flow rates, column length, column diameter and film thickness, precolumns, time measurement (min, s, scans), acquisition delay, different isothermal temperatures and different linear temperature ramps. The retention index might differ from the reference values owing to overloading effects, column contamination, reactive or decomposing compounds.

As can be seen in Figure 6.7, an *n*-alkane mixture from C10 to C18 was analysed on a specific GC-MS system and their retention times tabulated. As the retention index is defined for these compounds, if C10 = 1000, a linear correlation can be drawn between the retention time and retention index. This correlation can then be used for the determination of RI for the target compounds 1 and 2, calculated as 1100 and 1750 (Figure 6.7, right).

However, if the system undergoes a change, such as a longer column is installed, then this will result in different retention times. Rather than re-analysing all of the target compounds, only the hydrocarbons would need to be reanalysed to determine their new retention times. The retention index can now be used to predict the probable retention times for the two target compounds, as shown in Figure 6.8.

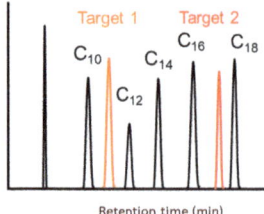
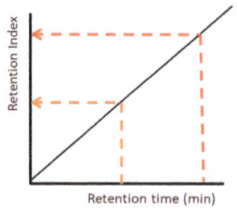

Name	Rt (min)	RI
C10	12.02	1000
C12	14.04	1200
C14	16.03	1400
C16	18.03	1600
C18	20.04	1800
Target 1	13.10	1100
Target 2	19.45	1750

Figure 6.7 A schematic diagram of a chromatogram of a standard containing a series of hydrocarbons and two target compounds (left); how retention time and linear retention index correlate (centre) and the tabulated data (right). Courtesy of Anthias Consulting Ltd.

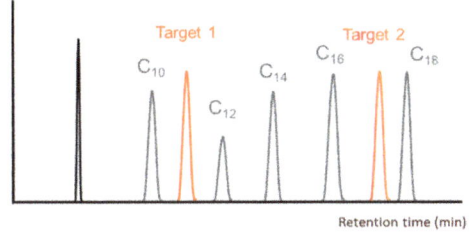

Name	RI	Rt (mins)
C10	1000	16.03
C12	1200	18.03
C14	1400	20.05
C16	1600	22.01
C18	1800	24.04
Target 1	1100	17.10
Target 2	1750	23.45

Figure 6.8　A schematic of a chromatogram showing how retention times have shifted (left) and the tabulated data (right) with the new retention times for the analysed hydrocarbons. The retention times for the target compounds 1 and 2 are calculated from their RI and the hydrocarbon new retention times. Courtesy of Anthias Consulting Ltd.

6.4　How Do I Identify Peaks?

Identification of peaks is usually achieved by library searching. However, it is important to obtain a "high quality spectrum" which will improve the match similarity and reduce the likelihood of mis-identification.

6.4.1　How Do I Extract a High-quality Mass Spectrum?

There are two parts to obtaining a cleaned-up spectrum, spectral averaging and background subtraction.

6.4.1.1　Spectral Averaging

The concentration of the analyte molecules changes across a peak, from low concentration up to a high concentration at the peak apex and then back to low again. With scanning instruments the concentration changes as the ion transmitting through the mass analyser to the detector changes m/z, which results in the ratios of the ions on the leading edge of the peak being different to those on the tailing edge of the peak, this is called peak skewing or tilting. This phenomenon is minimised by having at least 15–25 data points across the peak, the lower the number the greater the change in concentration between the data points. Spectral averaging across the peak reduces the influence of any spectral tilting or skewing, as one side cancels out the other. It is recommended to average the spectra at half to ¾ of the peak height on both sides of the peak, as shown in Figure 6.9.

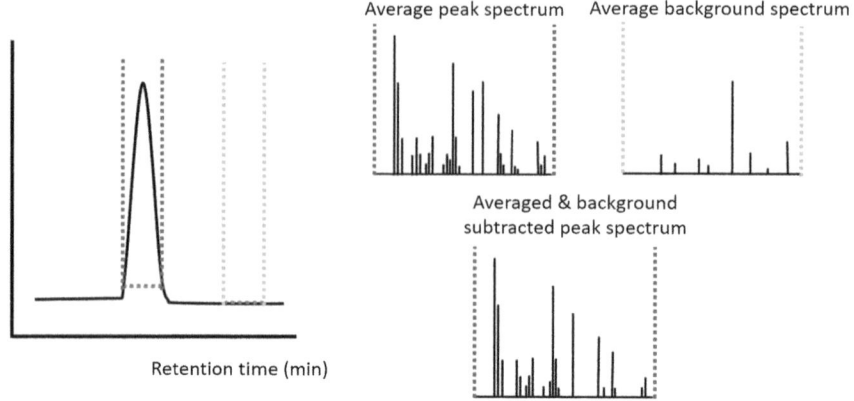

Average peak spectrum   Average background spectrum

Averaged & background
subtracted peak spectrum

Retention time (min)

Figure 6.9   Schematic diagram of peak averaging and background subtraction.

6.4.1.2 Background Subtraction

Background subtraction is another powerful tool that can be used to remove any erroneous ions in your spectrum that may not be part of your peak and could be present in the background but not from another closely eluting compound. For example, ions such as m/z 44 could be from CO_2 which would be present throughout your chromatogram, or m/z 281 could be a contribution from column bleed. The presence of these m/z ions in your spectra could lead to low similarity matches or even false peak identification. In most software, the average peak spectrum is taken, then the average or single point background spectrum is subsequently subtracted to produce a higher quality, averaged and background subtracted peak spectrum, as shown in Figure 6.9. Again, how to perform these actions is software-dependent.

6.4.1.3 Peak Purity

A feature in many GC-MS software applications is the ability to perform peak purity analyses, which is a good check to see if all the ions in an integrated peak actually belong to that peak. This is accomplished by looking at the spectrum of a peak, selecting several of the key ions and extracting them. This can also be performed manually. The analyst can then view the extracted ion chromatograms and see whether they all maximize at about the same time (for scanning instruments there may be an offset owing to spectral skewing) and they should be the same shape. If not, the peak is impure and some ions are from co-eluting peaks. As shown in Figure 6.10, the TIC peak at retention time 27.730 min consists of at least two peaks.

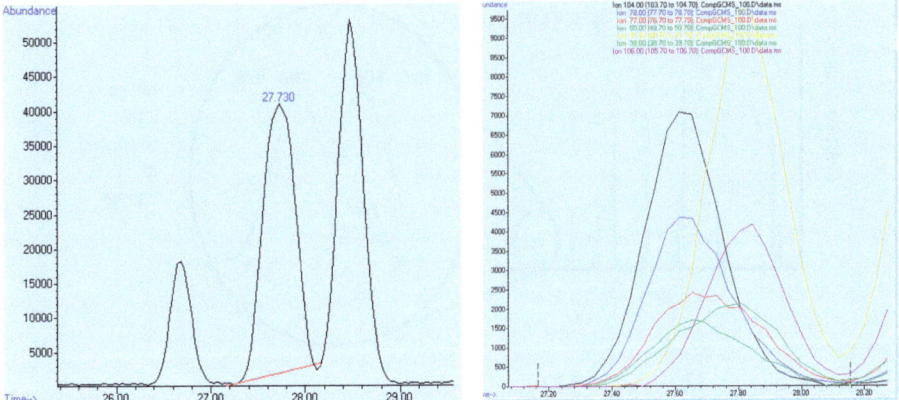

Figure 6.10 The integrated peak at Rt 27.730 in the left chromatogram has been subjected to the peak purity algorithm and the key ions have been extracted as EICs in the right chromatogram, showing that there are at least two compounds present. The screenshots were obtained from Agilent MSD ChemStation software. © Agilent Technologies, Inc. Reproduced with Permission, Courtesy of Agilent Technologies, Inc. The RSC accepts no liability for the accuracy or reproducibility of screenshots.

6.4.1.4 Mass Resolution

If a chromatogram does not have 100% chromatographic resolution of the peaks, mass or spectral resolution can be used as an additional separation method when a mass spectrometer is hyphenated to a GC (it is not possible with GC detectors).

Very small differences in the *m/z* can be used to differentiate coeluting peaks when a high resolution mass spectrometer is used. Mass resolution (*R*) is the ability to distinguish two ion peaks of slightly different *m/z* ratios (ΔM) as shown using eqn (6.1)

$$R = \frac{M1}{\Delta M}$$

(6.1)

for a specific *m/z* and height, for example:

- h at 10% in which peaks defined to be separated at h < 10% H;
- h at 50% or FWHM (full width at half maximum where 50%).

This is illustrated in Figure 6.11.

An ion from two different co-eluting analytes, even having the same nominal mass, is very unlikely to have the same exact mass to four

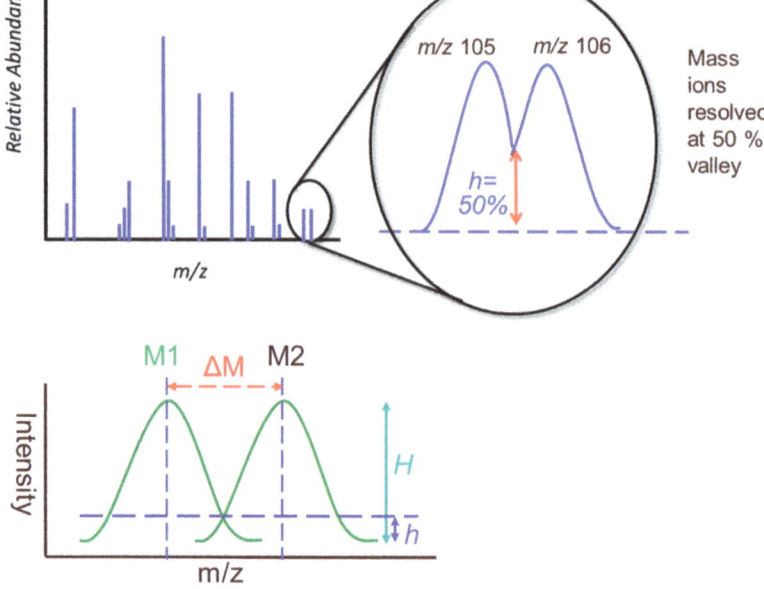

Figure 6.11   A mass spectrum showing mass resolution for *m/z* 105 and 106 resolved at a 50% valley, and below, a schematic diagram of how mass resolution can be calculated for mass *M1* and *M2* at 10% valley. Courtesy of Anthias Consulting Ltd.

decimal places, therefore a target analyte can be differentiated from, for example, a matrix peak with a similar retention time when using HRMS.

6.4.1.5   Spectral Resolution and Deconvolution

Peak purity is a great check for any coelutions of integrated peaks, however extracting a cleaned-up mass spectrum for identification, or finding small peaks underneath much larger peaks involves more complex software.

Target analytes only partially separating on a GC column can be identified and quantified by the differences in their mass fragmentation patterns. The exceptions are isomers, as they have the same or similar mass spectra and ions. If all of the ions in a mass spectrum belong to the same peak (analyte), then their concentration will increase and decrease at the same rate, after spectral deskewing of the data.

As discussed above, scanning instruments such as the quadrupole MS, allow ions of different *m/z* through the mass analyser and into the detector at slightly different times. If all of the ions acquired were overlaid, it would be noted that the ions, even if belonging to the same peak, would go up and down at very slightly different times, in the order that they were acquired, for example high to low mass or low to high mass depending on the manufacturer. Deskewing is the act of aligning all the ions to remove the acquisition time differences.

After deskewing, the ions belonging to the same peak will have the same apex and peak shape. Therefore, if an ion has a different apex or shape, it is very likely to come from a different analyte, the matrix or baseline noise that has a similar retention time on the analytical column.

The act of monitoring the rate of the rise and the fall of all the ions collected and then putting together a cleaned-up mass spectrum for each analyte, that is also library searchable, is called deconvolution, and is illustrated in Figure 6.12. Peaks that have the same apex and shape, for example two peaks that are totally co-eluting, cannot be deconvoluted and will result in a mixed mass spectrum that is difficult to interpret and will make the compounds difficult to correctly identify. This situation could arise owing to a lack of chromatographic resolution and therefore the GC separation conditions should be optimised.

There may also be two peaks which are slightly separated chromatographically but there may not be enough data points across the peaks for accurate determination of the apex, in which case the MS method must be optimised.

The use of deconvolution is useful in: fast GC analyses, in which total chromatographic resolution is not always achieved; in the analysis of samples with complex matrices, especially with large differences in the concentration of peaks, where small peaks of interest can often be masked by large, overloaded matrix components which cannot be chromatographically separated on a single column stationary phase; and for finding small peaks under the baseline.

For good deconvolution: the chromatographic peaks should be sharp, giving a good SNR and peak shape; and there must be enough data points across the peaks to enable the deconvolution of closely co-eluting peaks. In practice, 15 to 25 data points across the peak is optimal.

Deconvolution is included in some manufacturer's DA systems. AMDIS is a deconvolution program included with the NIST libraries

TIC and Spectrum   **Deconvoluted Peaks and Spectra**

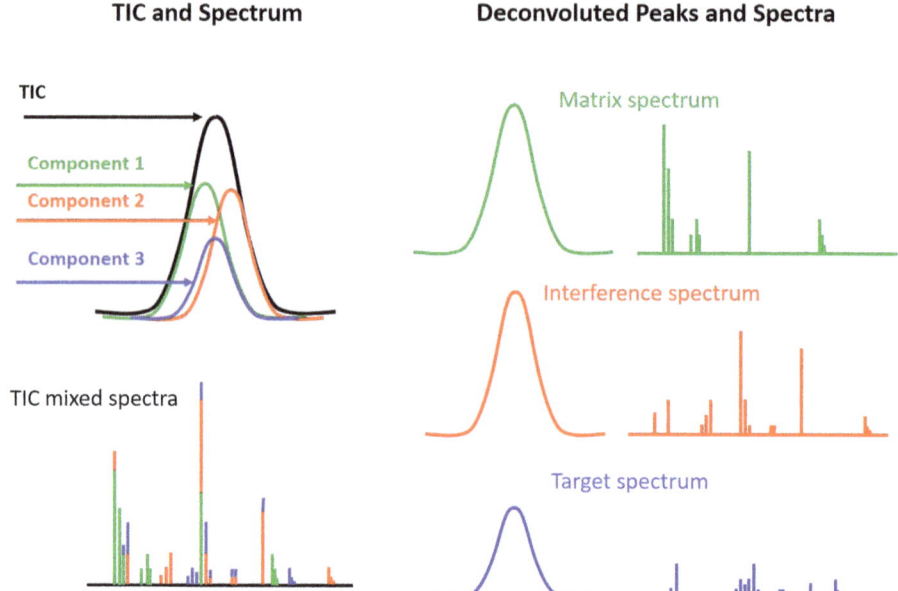

Figure 6.12   A schematic diagram of deconvolution extracting the matrix and interference spectra from the target spectrum. Courtesy of Anthias Consulting Ltd.

which can deconvolute data files from many manufacturer's software. It is very useful for nominal mass GC-MS instruments.

6.4.2   How Do I Perform a Library Search?

Library searching is a quick tool for the identification of chromatographic peaks. Library searching is a process to identify an unknown component, or to confirm the identity of a peak, by comparing its spectrum to reference spectra registered in mass spectral libraries. The "cleaned up" mass spectral pattern obtained acts as a compound fingerprint. Databases of mass spectra defined as a "mass spectral library" can be created by users or purchased from organisations such as NIST and Wiley. These libraries contain spectra of thousands of compounds, and all have been ionised using EI with 70 eV.

As shown in Figure 6.13, a comparison between the spectra at retention time 7.737 min in a data file was searched against the NIST14 database. Figure 6.13a shows a comparison between the searched spectrum (top) and the library spectrum (bottom). Figure 6.13b shows the top 10 hits with a top similarity match of 98%. Note that libraries can contain multiple spectra for a compound, all of

(a)

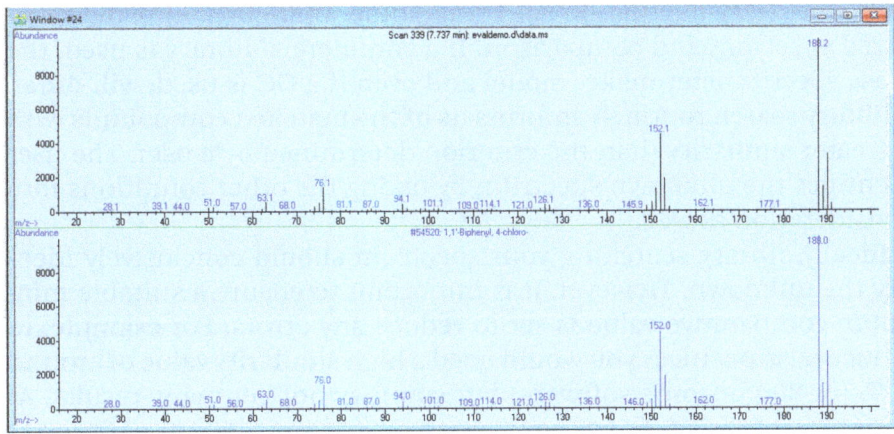

(b)

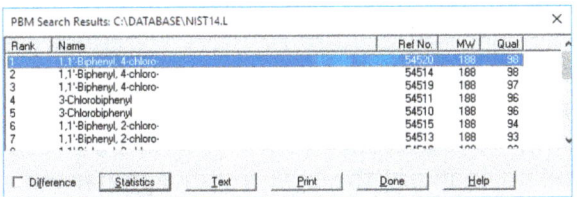

(C)

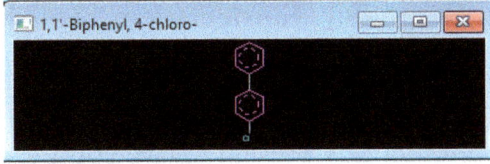

Figure 6.13 (a) Comparison between the searched spectrum (top) and the library spectrum (bottom); (b) top 10 hits with similarity matches; and (c) structure of the compound based on the hit selected. These images are obtained from Agilent MSD Chemstation software. © Agilent Technologies, Inc. Reproduced with Permission, Courtesy of Agilent Technologies, Inc. The RSC accepts no liability for the accuracy or reproducibility of screenshots.

which are very slightly different and therefore give slightly different similarity matches, as is seen here. Figure 6.13c shows the structure of the compound based on the library hit selected, if this information is included in the database.

In practice, exactly the same spectrum will not be obtained even though the same compound is measured. Many practical conditions such as how the instrument is used, other parameters in the measurement, sample condition, sample matrix, analyte concentration, plus statistical fluctuation all bring some ambiguity into the measured spectrum and comparison. If a commercial library is used, the mass spectrometer make, model and even if a GC is used, will differ. A library search routinely informs us of the matched compounds with a greater similarity than the criterion determined by a user. The user identifies the unknown spectrum by taking the other conditions into consideration as well.

Ideally, library searching your spectrum should conclusively identify the unknown. However, it is important to ensure a suitable minimum comparative value is set to reduce any errors. For example, to reduce false positives you would need a high similarity value of around 80% (or 800 in some software) but you may not get many results. At a lower value of 50%, you will observe many results but with lower matches and you may not correctly 'identify' the compound. However, if the peak is very tiny in a complex matrix then even low similarity matches may be correct. The user will need to be more vigilant with such matches and visually observe the comparative results.

The library comparative results will give a top ten similarity matches (or a number defined in your software criteria). It should be noted that, at times, the top hit result may not the best match against your analyte. The analysts' observation can often be more accurate in making the correct decision than the software search algorithm.

However, library searching can be limiting, considerations should include:

- Coeluting peaks: Are you sending the correct spectrum to search or could there be coeluting compounds present? Have you performed a peak purity test to ensure that the spectrum is pure?
- Is the unknown compound present in the library? There are many library databases with hundreds of thousands of spectra, however, you may be working with novel compounds that may not be present in these libraries.
- Do you have any known isomers? Other isomers may have the same mass spectrum as your compound and it can be very difficult to differentiate between them. One way to separate and differentiate between these compounds is through chromatographic resolution.
- Hydrocarbons, and other similar homologues series, can have a similar mass spectrum and the only way to differentiate between them would be the presence of any molecular ion, which if small in abundance can further diminish any library searching capabilities. This has been demonstrated further in Chapters 5 and 7.

- Results from a library search may not be conclusive for complete identification and you will then need to consider other techniques, such as NMR, UV-Vis and Fourier transform infrared spectroscopy (FTIR), to confirm the identity of the compound. This could be due to a low similarity match and/or the presence of isomers.

6.4.2.1 What Will Affect the Library Search Results?

There are several factors that can have an impact on the results that you obtain through library searching. These include:

- Software used: software from different manufacturers may use propriety or open source algorithms for library searching. These may produce slightly different results, especially with trace level compounds with poor spectral quality.
- Search algorithm: there are two main algorithms used to provide the similarity match:
  - o Forward search (FS): this is a process which evaluates the ions present in your spectrum and matches them against the ions in the library spectrum.
  - o Reverse search (RS): this is a process which evaluates the ions present in the library spectrum and matches them against the ions in your spectrum.

Depending on the manufacturer, both or either result may be presented. As there are two different processes applied, the results can also be distinctly different. For example, if there is a significant difference between the two search values, such as FS 899 and RS 701, this could indicate that ions are missing from your mass spectrum, for example is the mass range wide enough and is the molecular ion missing?

- Mass spectrum searched: some algorithms search all ions present in the mass spectrum, a threshold value should be applied to ignore low abundance ions (grass and noise). Other search algorithms only search the most abundant and unique masses, reducing library search time, and can give higher quality matches as the grass is always ignored. However, the user should be aware of how the algorithm searches for matches.
- Search strategy: the search strategy can speed-up library search time, for example looking at mass spectra for compounds under a

certain molecular weight. Also, when searching commercial libraries in which there are multiple spectra of the same compound,
selecting the option to give the top hit for each CAS number is very
useful to avoid replicate matches, as seen in Figure 6.13.

- A high-quality mass spectrum (averaged and background subtracted) should always be used. This is demonstrated in Figure
  6.9.

6.4.2.2   *What Are the Advantages of Creating a User Library?*

A user created library is a useful way to store regularly used spectra
for the purposes of library searching. Most GC-MS software has the
capabilities of creating a user library. Some of the benefits of a user
library are:

- Provides the fastest searches as it may be searching a database of
  100 s of compounds rather than 100 000 s.
- Provides best confidence with the highest similarity matches, as
  the spectra stored are obtained using your instrument.
- Known quality of reference spectra.
- Libraries and compound information can be edited and updated
  to include only relevant spectra. For example, a pesticide library
  which includes only spectra for pesticide compounds.
- Can include matrix peaks, which do not need to be identified but
  are usually present.

6.4.3   My Mass Spectrum Is Not in the Library, How Do I Determine What It Is?

Today's libraries are very large, however not all organic compounds
are included within them. If not, a high quality, cleaned-up mass spectrum, obtained as described in this chapter, can be interpreted. Mass
spectral interpretation is covered in detail in Chapter 7.

7 Basic Aspects of Mass Spectra Interpretation

7.1 What Information Does the Molecular Ion in EI–MS [M]$^{+\bullet}$ Give Me?

7.1.1 What Is a Molecular Ion?

We have seen that the molecules can be ionized using different methods. The nature of the molecular ion is dependent on the way it was formed. In electron (impact) ionization (EI)-MS the molecular ion [M]$^{+\bullet}$ is generated by loss of an electron (Section 5.4.1, Chapter 5). The respective molecular ion must have an *odd* number of electrons and is a radical cation = OE$^{+\bullet}$ (OE$^{+\bullet}$ odd electron number ion). In chemical ionisation (CI) however, the choice of the reagent gas determines whether the molecular ion is an OE$^{+\bullet}$ type ion, generated *via* charge exchange, or is produced *via* protonation (see Section 5.4.2, Chapter 5). In the latter process molecular ions are closed shell species of the type [M + H]$^+$ and have an even electron number (EE$^+$ ions). It is therefore important to note that a correct interpretation of a mass spectrum starts with the correct understanding of the ionisation method used. The nature of the molecular ion is especially important for the correct determination of molecular compositions on the basis of accurate ion mass measurements and is also crucial for the correct understanding of fragmentation patterns.

Gas Chromatography-Mass Spectrometry: How Do I Get the Best Results?
By Diane Turner, Mathias Schäfer, Steven Lancaster, Imran Janmohamed, Anthony Gachanja and Jason Creasey
© Diane C. Turner, Mathias Schäfer, Steven Lancaster, Imran Janmohamed, Anthony Gachanja and Jason Creasey 2020
Published by the Royal Society of Chemistry, www.rsc.org

In the following discussion we want to concentrate on the interpretation of EI-MS spectra as those are by far the most important ones in GC-MS. As discussed in Section 5.4.1 in Chapter 5 molecular ions $[M]^{+\bullet}$ generated by EI-MS can either be stable or unstable and the relative intensity of the molecular ions relies strongly on the nature of the analyte. Aside from that, the elemental composition is reflected in the isotopic signal pattern of the molecular ions $[M]^{+\bullet}$ found in the EI mass spectra, which is an important piece of information on the identity of the ion of interest.

7.1.1.1 What Kind of Information Does the Molecular Ion Provide?

In EI-MS the molecular ion of a typical organic molecule not containing nitrogen will have an *even mass-to-charge* ratio *m/z*. This relationship is fundamental to the correct interpretation of the mass spectrum. For example, the molecular ion of propane $[C_3H_8]^{+\bullet}$ is found at *m/z* 44, the one of benzene $[C_6H_6]^{+\bullet}$ at *m/z* 78 and the one for tri-phenyl phosphine $[(C_6H_5)_3P]^{+\bullet}$ at *m/z* 262. Organic compounds $(C_uH_vN_wO_xHal_yS_z)$ with an uneven number of nitrogens have an uneven nominal molecular mass (see Section 5.2.5 on the nominal ion mass). This finding is the result of the fact that the elements typically found in organic molecules $(C_uH_vN_wO_xHal_yS_zP_n)$ have either an uneven number of valences (*e.g.* hydrogen = 1) and also an uneven atomic mass (^1H A_r = 1 u) *or* an even valence number such as carbon = 4 and an even atomic mass (^{12}C A_r = 12u). However, nitrogen is an exception here as it has an uneven number of valences (= 3) but an even atomic mass (^{14}N A_r = 14u). This is the foundation of the so-called *nitrogen rule* that is of ample importance for the first evaluation of the *m/z* ratio of a given molecular ion to probe a hypothetical composition.

7.1.2 What Is the Nitrogen Rule?

According to the nitrogen rule the *m/z* value of a molecular ion provides valuable information on the number of nitrogens present.[1-3] The nominal mass of the molecular ion of ethylamine $[C_2H_5NH_2]^{+\bullet}$ is found at *m/z* 45 (one nitrogen gives an uneven mass), the one for strychnine $[C_{21}H_{22}N_2O_2]^{+\bullet}$ at *m/z* 334 (2 nitrogens gives an even mass), $[C_{17}H_{35}COOH]^{+\bullet}$ at *m/z* 284 (no nitrogen gives an even mass). In Figure 7.1 the EI mass spectrum of *i*-propyl-methyl-pentylamine is shown to illustrate this. The molecular ion is found at *m/z* 143 confirming the presence of one nitrogen in the composition $(C_9H_{21}N)$. However, all relevant fragment ions found at *m/z* 128,

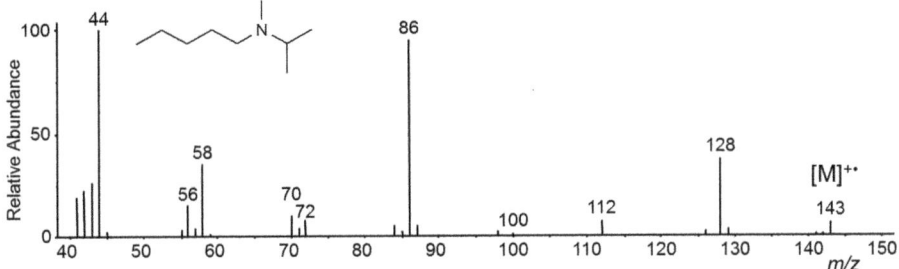

Figure 7.1 EI-mass spectrum of methyl-pentyl-i-propyl-amine with an uneven molecular mass according to the nitrogen rule. The molecular ion of i-propyl-methyl-pentylamine is found at *m/z* 143 in the EI mass spectrum. Spectrum taken from the NIST Mass Spectral Library, Version 2.0f, 2008. Reproduced with permission.

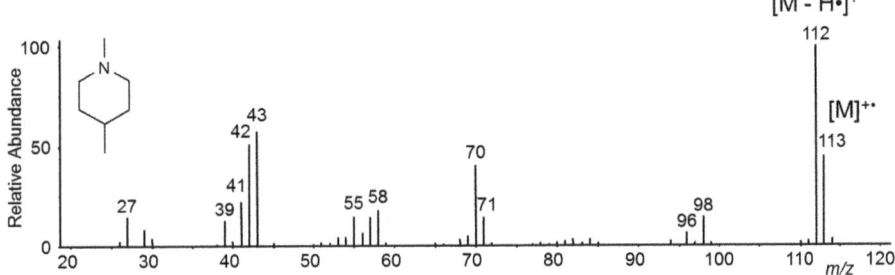

Figure 7.2 EI-mass spectrum of 1,4-dimethylpiperidine ($C_7H_{15}N$). Spectrum taken from the NIST Mass Spectral Library, Version 2.0f, 2008. Reproduced with permission.

86, 58, and 44 have even *m/z* values as they represent *even electron number* fragment ions EE^+ formed by the loss of a radical from the open shell molecular ion $[M]^{+\bullet} \rightarrow [EE]^+ + R^{\bullet}$ (this will be discussed in detail below). Keep in mind that in CI-MS in the case of protonated, closed shell EE^+ type molecular ions $[M + H]^+$ the respective *m/z* values will be shifted by 1!

Mass spectra conventionally show the *m/z* scale on the *x*-axis (note: sometimes *m/z* is also called Th/Thompson), which is in practice the mass of the ionised species, as in EI-MS usually the charge *z* will be 1, whilst the *y*-axis shows the relative intensity/abundance of the ions normalised to 100.

However, it is important to have a close look at the signal pattern that is found at the highest end of the *m/z* scale in the mass spectrum. The EI-mass spectrum of 1,4-dimethylpiperidine shown in Figure 7.2 serves as a good example for that.

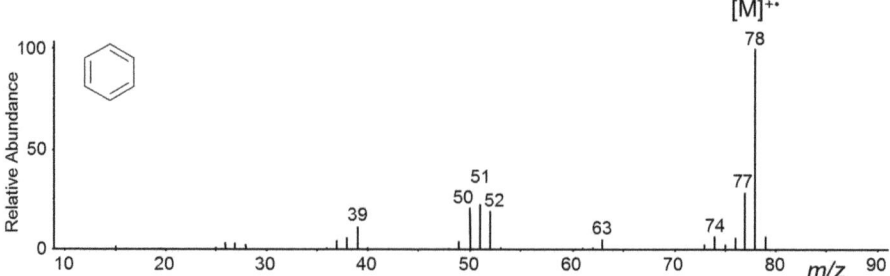

Figure 7.3   EI mass spectra of benzene with a strong molecular ion at *m/z* 78 and a few weak fragment ions. Spectrum taken from the NIST Mass Spectral Library, Version 2.0f, 2008. Reproduced with permission.

Here, the signal with the highest relative abundance in the spectrum (the so called base peak) is found at *m/z* 112 which refers to a strong hydrogen elimination process. The actual molecular ion of the analyte is found at *m/z* 113. This makes sense according to the nitrogen rule and the composition of the ion. The fragmentation process responsible for that hydrogen loss (1 *u* = H˙) is called an *alpha*-cleavage as the α-bond relative to the location of the unpaired electron and the positive charge is cleaved in this process. Such *alpha*-cleavages are typical for amines and other organic molecules and will be discussed below in Section 7.2.6.

7.1.3   Double Bond-equivalents

The number of double bonds and/or the number of rings, known as the double bond-equivalents (DBE) is a helpful criterion for the interpretation of an EI-MS spectrum of an unknown compound, as the EI-spectra of certain classes of organic compounds show characteristic features. For example, the EI-spectra of saturated hydrocarbons differ strongly from those of unsaturated or aromatic compounds. EI-mass spectra with a strong molecular ion and a few weak fragment ions point towards un-substituted aromatic or hetero-aromatic analytes (*e.g.* EI mass spectrum of benzene, see Figure 7.3; or naphthalene or quinoline).

7.1.3.1   *What Can I Learn From the Profile of the Mass Spectrum?*

EI-mass spectra with a strong molecular ion plus a few strong fragments are characteristic of substituted aromatic of hetero aromatic analytes such as benzophenone for example (Figure 7.4).

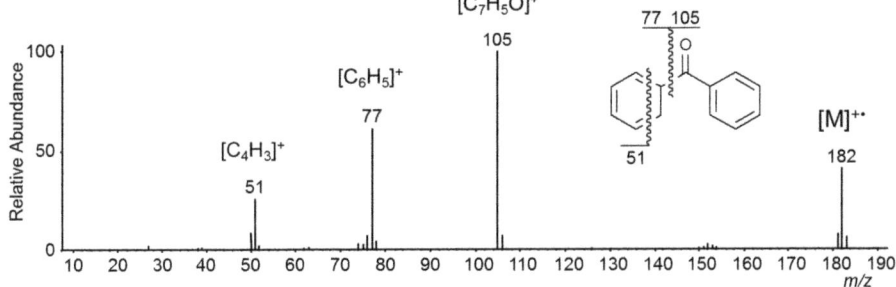

Figure 7.4 EI mass spectrum of benzophenone with a strong molecular ion at *m/z* 182 and a few strong fragment ions. The fragmentation reactions leading to the formation of the significant ions at *m/z* 105, 77 and 51 are explained in Scheme 7.20 in Section 7.2.8. Spectrum taken from the NIST Mass Spectral Library, Version 2.0f, 2008. Reproduced with permission.

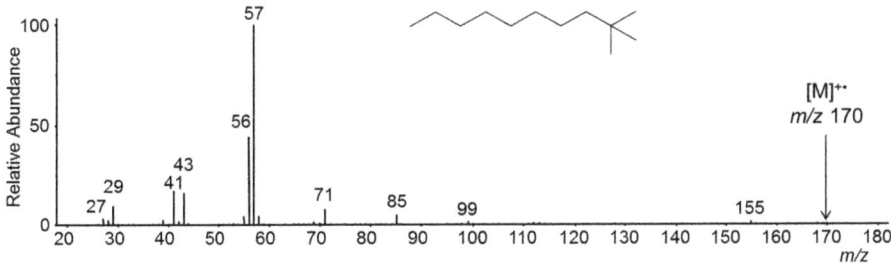

Figure 7.5 EI mass spectrum of 2,2-dimethyldecane in which the molecular ion is not observed. The formation of the very stable tertiary butyl fragment ion [(CH$_3$)$_3$C]$^+$ at *m/z* 57 is the prominent fragmentation pathway and hence, the molecular ions are completely depleted. The fragment ion at *m/z* 56 relates to the radical cation of methylpropene [(CH$_3$)$_2$C=CH$_2$]$^{+•}$ formed by a rearrangement including a hydrogen shift reaction and the loss of the respective octane as illustrated in Scheme 7.6 in Section 7.2.4. Spectrum taken from the NIST Mass Spectral Library, Version 2.0f, 2008. Reproduced with permission.

EI-mass spectra with a weak molecular ion and a series of fragments with Δm = 14 u = methylene moieties –CH$_2$– are characteristic for alkanes and alkenes, possibly with terminal substituents, for example, octane, octan-1-ol (compare with Figures 5.8 and 5.15, Chapter 5). EI-mass spectra with a low abundance molecular ion and a few strong fragments are often found in EI-mass spectra of branched alkanes (Figure 7.5), alkenes, ketones, esters, di and tri-alkylamines and ethers.

7.1.3.2   How Do I Calculate the DBE?

The number of DBE (number of rings + number of double bonds; triple-bonds count as 2 double bonds) of a given neutral compound with the formula $C_cH_hN_nO_o$ is calculated on the basis of the valence rules using:

$$\text{DBE} = c - \frac{1}{2}h + \frac{1}{2}n + 1 \tag{7.1}$$

The calculation can be extended for halogens, which are treated like monovalent hydrogen and for sulphur, which is treated like oxygen, as bivalent elements are cancelled out.

For example, the DBE of heptane C_7H_{16} confirms the saturation $\text{DBE}_{heptane} = 7 - 8 + 1 = 0$. The DBE of toluene C_7H_8 is delivers according to $\text{DBE}_{toluene} = 7 - 4 + 1 = 4$, indicating three pairs of π-electrons = double bonds and the ring system.[1,2]

7.1.4   What Information Does the Isotopic Patterns of the Molecular Ion [M]$^{+\bullet}$ in EI-Mass Spectra Give Me?

Carbon has two naturally occurring stable isotopes that are found with a characteristic isotopic abundance: $^{12}_6C$ (the most abundant isotope is normalised to 100%) and $^{13}_6C$ with 1.08%. In Table 5.1 (Chapter 5 Section 5.2.3) the isotopic composition of common elements is listed and the importance of this piece of information becomes clear when it comes to the examination and interpretation of signal patterns of molecular ions [M]$^{+\bullet}$ in mass spectra.

Figure 7.6 illustrates this for a hypothetical C_{10} cluster with a nominal mass of 120 u. The $[C_{10}]^{+\bullet}$ molecular ion is found at m/z 120 and a $[M + 1]^{+\bullet}$ ion signal with an intensity of 11% completes the isotopic pattern. The $[M + 1]^{+\bullet}$ signal reflects the natural abundance of the ^{13}C isotope of carbon (1.1%). We can translate this to simple hydrocarbon examples: methane CH_4 has a nominal mass of 16 u and a molecular ion $[M]^{+\bullet}$ at m/z 16 (100%) and the $[M + 1]^{+\bullet}$ signal (belonging to $^{13}CH_4$) has an intensity of 1.1%. The same holds true for pentane C_5H_{12} with a molecular ion $[M]^{+\bullet}$ at m/z 72 (100%) and a $[M + 1]^{+\bullet}$ signal at m/z 73 of $^{12}C_4{}^{13}CH_{12}$ with an intensity of 5.5% (see also Figure 7.4 for comparison). The EI mass spectrum of benzene (C_6H_6) is shown in Figure 7.3. The nominal mass of the molecular ion $[M]^{+\bullet}$ is found at m/z 78. According to the natural abundance of ^{13}C the $[M + 1]^{+\bullet}$ ion has an intensity of $6 \times 1.1\% \approx 6.6\%$ of $[M]^{+\bullet}$.

In the presence of metal or halogen atoms in the composition of an analyte under investigation, the respective isotopic distribution

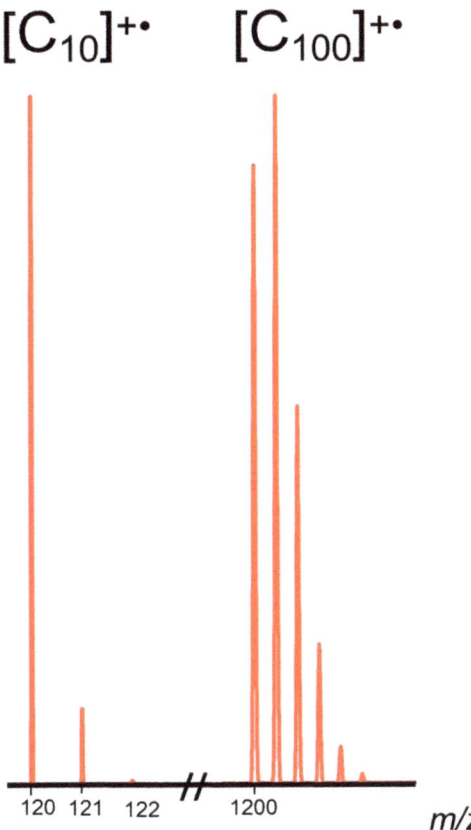

Figure 7.6   EI-MS molecular ion region of a hypothetical C_{10} cluster with a molecular mass of 120 u and a C_{100} cluster at m/z 1200. The nominal mass of the molecular ion $[M]^{+\bullet}$ of C_{10} is found at m/z 120 with a normalised intensity of 100%. According to the natural abundance of ^{13}C the $[M + 1]^{+\bullet}$ ion has an intensity of $10 \times 1.1\% = 11\%$. The signal distribution of the $[C_{100}]^{+\bullet}$ molecular ion is discussed below (see Section 7.1.4.2 on *large* molecular ions). In large molecular ions the signal of the $[M + 1]^{+\bullet}$ ion (here $^{12}C_{99}{}^{13}C$ at m/z 1201) is more abundant than the signal of $[M]^{+\bullet}$ (here: $^{12}C_{100}$ at m/z 1200, which is by definition the nominal mass). The isotopic distribution was computed using the Universal Mass Calculator (UMC) (Version 3.7.2.58) ©Matthias Letzel University Münster, Germany.

of stable isotopes must be found in the signal distribution of the respective molecular ion $[M]^{+\bullet}$. This is illustrated below in the EI-mass spectra of chloro- and bromo-benzene (Figure 7.7). It is highly recommended to probe any structural hypothesis with a theoretical signal distribution of the respective molecular ion to avoid incorrect assignments.

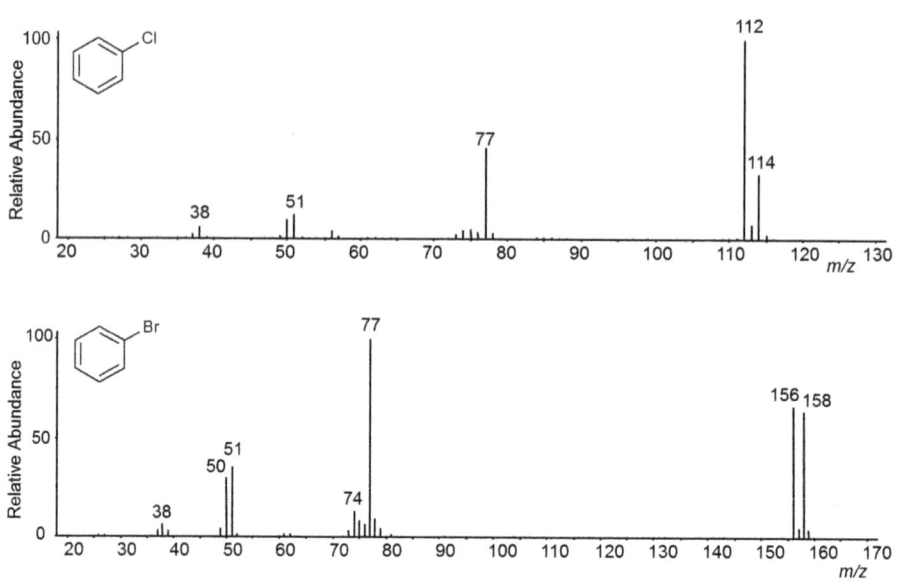

Figure 7.7 EI-mass spectra of chloro- and bromo-benzene. The characteristic isotopic distribution of the halogens is found in the signal patterns of the respective molecular ions [M]$^{+•}$. The nominal mass of [C$_6$H$_5$Cl]$^{+•}$ is found at *m/z* 112 ([^{12}C$_6$^{1}H$_5$^{35}Cl]$^{+•}$) and the contribution of the ^{37}Cl isotope of chlorine is obvious in the [M + 2]$^{+•}$ signal at *m/z* 114 with 32.4% ([^{12}C$_6$^{1}H$_5$^{37}Cl]$^{+•}$). A similar isotope analysis explains the signal pattern and the signal intensities of the bromo-benzene molecular ion: [^{12}C$_6$^{1}H$_5$^{79}Br]$^{+•}$ at *m/z* 156 and [^{12}C$_6$^{1}H$_5$^{81}Br]$^{+•}$ at *m/z* 158 with (^{79}Br 100% and ^{81}Br 97.5%; see Table 5.1 in Section 5.2.3, Chapter 5). Spectra taken from the NIST Mass Spectral Library, Version 2.0f, 2008. Reproduced with permission.

7.1.4.1   What Is the Probability of Having Two Carbon-13 Atoms in the Molecule?

For molecules with more than five carbons the natural abundance of the ^{13}C isotope leads to a significant probability of signals at [M + 2]$^{+•}$ resulting from **two** ^{13}C atoms in the molecule. The probability of having two ^{13}C carbon isotopes in a molecule and also the relative intensity of the [M + 2]$^{+•}$ signal in the EI-mass spectrum can be approximated using the following equation:

$$\% \, [M + 2]^{+•} = (\% \, [M + 1]^{+•})^2/200 \tag{7.2}$$

With this simple relationship the intensity of the [M + 2]$^{+•}$ signal at *m/z* 122 in Figure 7.6 can be calculated as: $(11)^2/200 = 0.6\%$. See also

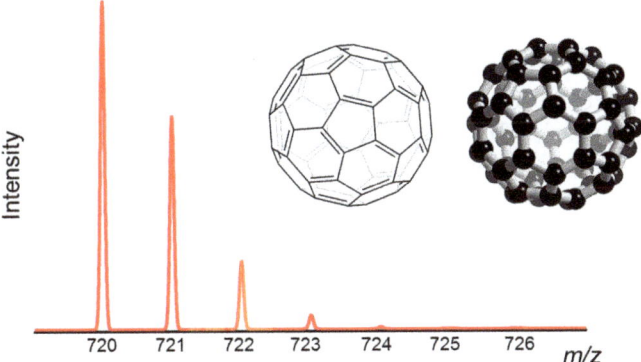

Figure 7.8   EI-mass spectrum section showing the *m/z* range of the molecular ion of the football-shaped C_{60} fullerene. The molecular mass of $[^{12}C_{60}]^{+\bullet}$ is found at *m/z* 720 (nominal mass with a normalised intensity of 100%). The relative intensity of the $[M + 1]^{+\bullet}$ signal at *m/z* 721 is $60 \times 1.1\% = 66\%$ and the one for the $[M + 2]^{+\bullet}$ signal at *m/z* 722 is calculated to be $(66)^2/200 \approx 22\%$. C_{60} illustrations are taken from the public domain. The isotopic distribution was computed using the UMC (Version 3.7.2.58) ©Matthias Letzel University Münster, Germany.

Figure 7.8 which shows the *m/z* range of the EI mass spectrum with the molecular ion of the allotropic carbon C_{60} fullerene.

7.1.4.2   Isotopic Contributions to Signal Patterns: Large Molecular Ions

As discussed above, carbon consists of 98.89% of the isotope ^{12}C and of 1.1% of ^{13}C. Hence, in molecules with 90 or more carbon atoms the probability of finding a pure $^{12}C_{90}$ molecule is lower than one with at least one ^{13}C isotope, that is, $^{12}C_{89}{}^{13}C$. Consequently, the signal for $^{12}C_{89}{}^{13}C$ at *m/z* 1201 is more abundant than the signal of $^{12}C_{90}$ at *m/z* 1200, which is by definition the nominal mass. In Figure 7.6 the *m/z*-range of the molecular ion $[C_{100}]^{+\bullet}$ at *m/z* 1200 is depicted. For the C_{100} carbon cluster the signal intensity of the nominal mass $[^{12}C_{100}]^{+\bullet}$ is lower than the signal intensity of the $[M + 1]^{+\bullet}$ ion: $100 \times 1.1 = 110\%$.

7.2   What Information Do the EI-MS Fragment Ions Give Me?

What are fragment ions? How are they formed? Are fragmentation processes predictable? Are there characteristic fragmentation pathways for classes of organic compounds?

7.2.1 What Are Localised Charges?

The considerations that lead to the so called *concept of localised charges* can be rationalised upon inspection of the EI mass spectra of 5α-pregnane and the directly related 20-dimethylamino-5α-pregnane. In the EI mass spectrum of the saturated steroid 5α-pregnane a large number of different fragment ions are observed, which highlight the similar stability of the respective C–C sigma bonds leading to many competitive fragmentation pathways. In contrast, in the EI mass spectrum of 20-dimethylamino-5α-pregnane only a single fragment ion at *m/z* 72 is detected which is responsible for more than 90% of the TIC (Figure 7.9). As the steroid skeleton of the 5α-pregnane system remains untouched in the latter analyte the dimethyl amino

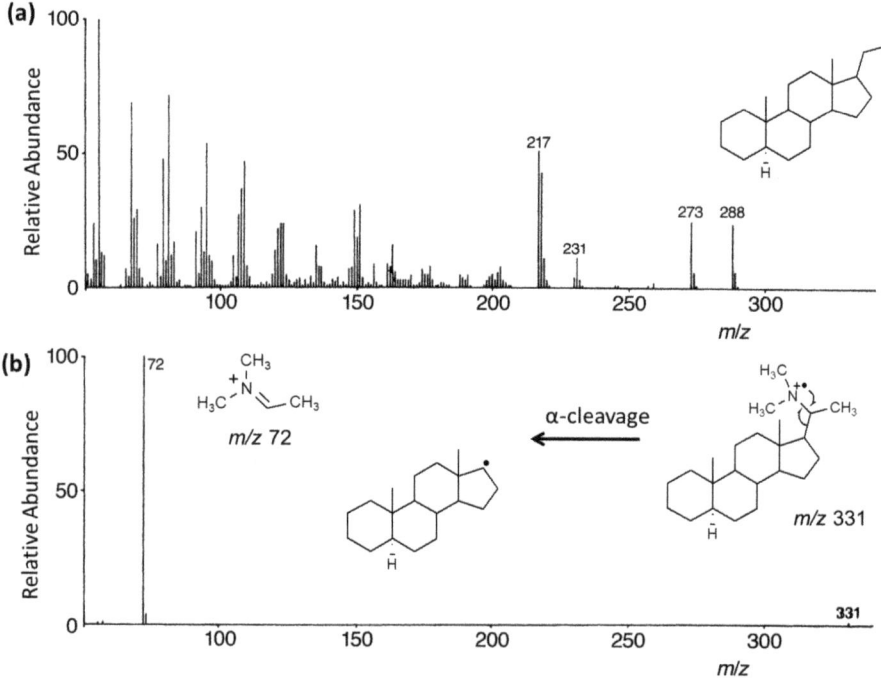

Figure 7.9 EI mass spectra of 5α-pregnane (a) and 20-dimethylamino-5α-pregnane (b). The free electron pair of the amine functionality is clearly the most favoured ionisation site. The radical cationic site at the amine nitrogen triggers the alpha-cleavage that provides a straight forward explanation for the exclusive formation of the ammonium fragment ion $[C_4H_{10}N]^+$ at *m/z* 72 from the respective molecular ion at *m/z* 331 (Schemes 7.10 and 7.11). Spectra reproduced from ref. 4 with permission from John Wiley and Sons, Copyright © 2003 John Wiley & Sons, Ltd.

functionality must be responsible for the dramatic change in the EI-MS behaviour. It is reasonable to assume that the electronically excited molecules expel an electron from their highest occupied molecular orbital (HOMO), in which the ionisation energy (IE) is supposed to be minimal (Figures 5.10a and 5.12, Chapter 5). The HOMO is typically a non-binding free electron pair of hetero-atoms or π-systems, if available. Directly after formation of the molecular ion $[M]^{+\bullet}$ the unpaired electron resides where the positive charge is located. In the case of 20-dimethylamino-5α-pregnane the free electron pair of the tertiary amine functionality is exclusively ionised and the sigma bond skeleton of the steroid ring system remains intact. The excess energy of the molecular ion $[M]^{+\bullet}$ received in the ionisation process allows further fragmentation reactions to occur (Figure 5.10b and 5.12, Chapter 5). The strong negative inductive effect of the radical cationic site in the molecular ion $[M]^{+\bullet}$ weakens the neighbouring bonds and fragmentation reactions of adjacent bonds are likely. In the case of 20-dimethylamino-5α-pregnane the radical cationic site at the dimethyl amino functionality leads to the exclusive the alpha-cleavage reaction (see inset in Figure 7.9). These considerations on the fundamentals of the charge/and radical site driven fragmentation and rearrangement reactions, which are discussed below (Section 7.2.3 and further on), build the framework of the *concept of localised charges.*[2,5]

7.2.2   How Do I Correctly Discuss MS Fragmentation Processes?

Radical cationic molecular ions are written in squared brackets with the '+' and the radical dot '''' as seen below, *for example* the molecular ion of acetone $[CH_3COCH_3]^{+\bullet}$

In some cases the structures shown in squared brackets can be minimised, as shown in Scheme 7.1.

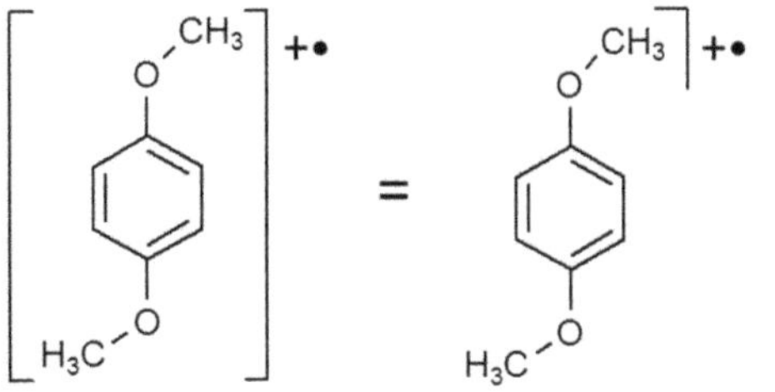

Scheme 7.1

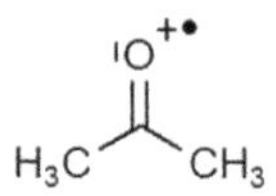

$$\underset{H_3C \diagdown \diagup CH_3}{\overset{\overset{+\bullet}{|O|}}{\big\|}}$$

Scheme 7.2

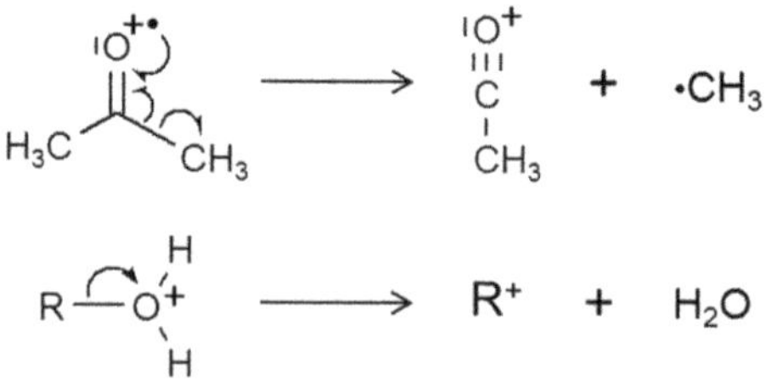

$$\underset{H_3C \diagdown \diagup CH_3}{\overset{\overset{+\bullet}{|O|}}{\big\|}} \longrightarrow \underset{CH_3}{\overset{\overset{+}{|O|}}{\underset{|}{\overset{|||}{C}}}} + \cdot CH_3$$

$$\underset{\underset{H}{|}}{R \diagdown \overset{\overset{H}{|}}{\underset{}{O^+}}} \longrightarrow R^+ + H_2O$$

Scheme 7.3

According to the *concept of localised charges* the radical cationic site in the molecule can be identified as illustrated in Scheme 7.2.

The discussion of fragmentation and rearrangement mechanisms necessitates the definition of the bond cleavage processes (Scheme 7.3). A homolytic cleavage which is equal to the migration of one electron is symbolized with a half arrow ⌒. A heterolytic cleavage which is equal to the migration of two electrons is symbolized with a complete arrow ⌒. The (alpha) α-cleavage of acetone is therefore written with a set of half arrows, whereas the loss of water from a protonated alcohol involves the shift of two electrons (full arrow).

In some cases bond cleavage processes are not written in full and are abbreviated. The number below or above the straight line is the *m/z* ratio of the charged ion found in the spectrum and the sinuous line, as shown in Scheme 7.4 (this can also be a straight line, see Figure 7.4 for example), denotes the cleaved bond.

7.2.3 Typical Fragmentation and Rearrangement Reactions in EI-MS

We have seen in the previous chapters (Chapter 5, Section 5.4) that in routine EI-MS 70 eV electrons are used which will transfer a substantial amount of extra energy on the primarily formed analyte molecular ions $[M]^{+\bullet}$. The molecular ions are highly activated and organic analytes show characteristic fragmentation reactions which provide detailed insights into the molecular structure. In addition to the identification of molecules the distinction of constitutional isomers is often possible. According to the *concept of localised charges* (see Section 7.2.2) the fragmentation reactions can be correlated directly to

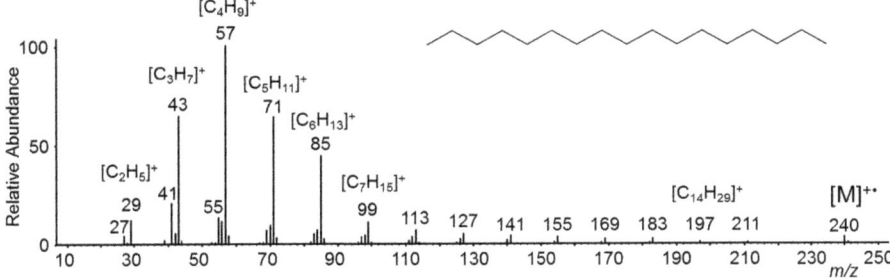

Scheme 7.4

Figure 7.10 EI-MS of heptadecane shows an undisturbed distribution of abundant fragment ions in the low m/z range with a maximum around 3–5 carbons in $[C_nH_{2n+1}]^+$ ions after the loss of a radical. The fragment ion signals exhibit mass differences of 14 u, which corresponds to the mass of a methylene unit (CH_2). Spectrum taken from the NIST Mass Spectral Library, Version 2.0f, 2008. Reproduced with permission.

functional group characteristics which in turn allow understanding of the fragmentation reactions (benzylic-, allylic-, α-cleavages and also rearrangement reactions; see Scheme 7.10 for example). In the following we want to look at exemplary fragmentation and rearrangement reactions that represent the EI-MS spectra of different organic classes of analytes and therefore correlate directly to the molecular structure allowing their identification.

7.2.4 Fragmentation of Unactivated C–C Sigma Bonds: $OE^{+\bullet} \rightarrow EE^+ + R^\bullet$

In saturated hydrocarbons only sigma bonds are available for the loss of an electron in the EI ionisation process. The ionised bond is substantially destabilised and prolonged and this bond is most likely to be broken in the primary fragmentation process of the molecular ion. The loss of the respective radical will deliver an EE^+ type fragment ion. For *n*-alkane molecules the predominant EE^+ type fragment ions have the general composition of $[C_nH_{2n+1}]^+$. The fragment ion intensity has a maximum at the respective ions with 3–4 carbons (see the EI mass spectrum of heptadecane in Figure 7.10). In the case of branched alkanes the distribution of

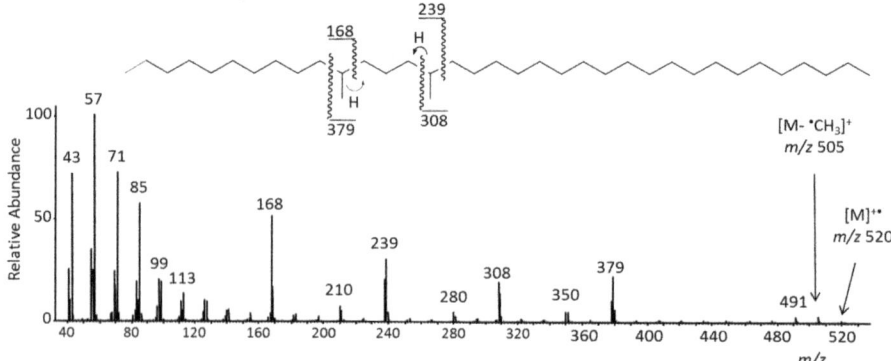

Figure 7.11 EI mass spectrum of the branched 11,15-dimethyl-pentatriacontane ($C_{37}H_{76}$) with the molecular ion at m/z 520. The prominent fragment ions at an uneven mass at m/z 379 and 239 refer to hydrocarbon EE$^+$ type (secondary) carbocations with the formula $[C_nH_{2n+1}]^+$. They are formed by fragmentation at the branching points of the molecule by loss of the respective primary radicals. The OE$^{+\bullet}$ type fragment ions $[C_nH_{2n}]^{+\bullet}$ at an even mass at m/z 168 and 308 also result from fragmentation at these branching points, but involve hydrogen shifts as indicated in the inset (see also Scheme 7.6). Spectrum taken from the NIST Mass Spectral Library, Version 2.0f, 2008. Reproduced with permission.

fragment ion intensities is disturbed owing to the elevated stability of the carbocations that are a result of the fragmentation at the branching points.

The EI mass spectrum of a long chained alkane is included in Figure 7.11, in which four prominent fragment ion signals indicate the di-methyl substitution. This observation follows the straight forward explanation that the thermodynamically most stable fragment ions will be formed upon the cleavage of the weakest bond. Consequently, the *preferred* fragmentation takes place at highly substituted carbons (tertiary > secondary > primary) delivering the most stable carbocations, by virtue of the positive inductive effect of the alkyl substituents at the respective carbon (Scheme 7.5 and Figure 7.13). A striking example of this consideration is found in the EI-MS fragmentation pattern of 2,2-dimethyldecane (Figure 7.5). The EI mass spectrum is dominated by a very strong $[(CH_3)_3C]^+$ fragment ion at m/z 57 and the molecular ion is not observed, as the formation of the former is much more favoured. The fragmentation can be rationalised on the basis of the inspection of the IE values of the competing fragments (Schemes 7.6 and 7.7 and Figure 7.13).[1–3]

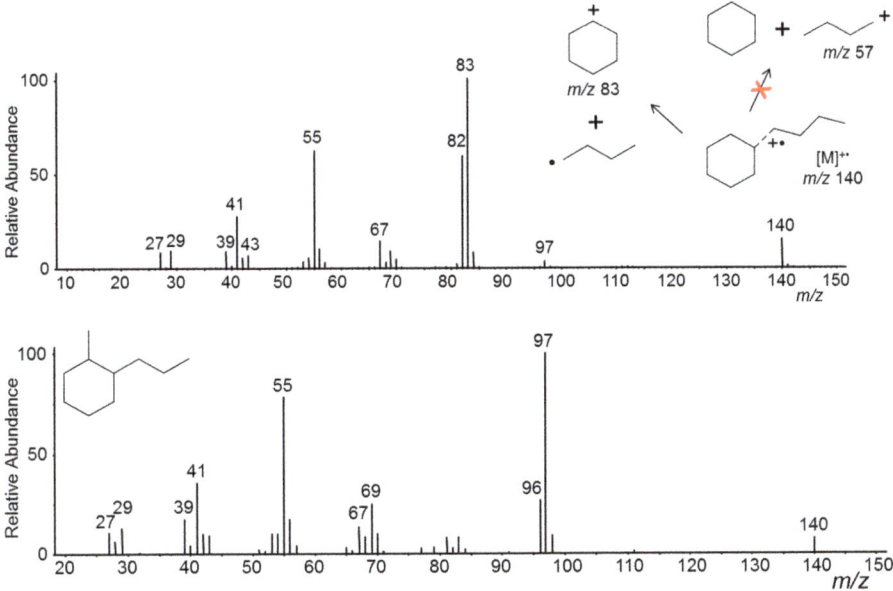

Figure 7.12 EI mass spectra of two substituted isomeric cyclohexane derivatives with the molecular mass of 140 u. The butyl cyclohexane delivers a strong fragment at *m/z* 83 in the EI mass spectrum motivated by the relatively high stability of a secondary carbenium ion *versus* a primary carbenium ion (see inset on the right). The dominating fragmentation reaction of the butyl cyclohexane molecular ion leads predominantly to the secondary carbocation at *m/z* 83 as the spectrum documents. The EI mass spectrum of 1-methyl-2-propylcyclohexane shows a similar fragmentation pathway with the formation of the mass shifted ion at *m/z* 97, by the loss of a propyl radical, offering a distinct differentiation between the two isomers. Spectra taken from the NIST Mass Spectral Library, Version 2.0f, 2008. Reproduced with permission.

$$R_3C^{+\cdot}R \rightarrow R_3C^+ + R^\cdot$$

Substituted alicyclic alkanes and steroids show a similar regioselectivity in the fragmentation behaviour for the same reason as the branched alkanes. The EI mass spectrum of 5α-pregnane (Figure 7.9a) and those of the substituted cyclohexane isomers serve as examples in this respect (Figure 7.12 and Scheme 7.5). An unsubstituted cycloalkane shows a similar EI mass spectrum as an alkene, as the initial loss of a radical delivers OE$^{+\cdot}$ type ion series (Figure 7.14).

After close inspection of the spectra and schemes detailed above (Figures 7.5, 7.11 and 7.12 and Schemes 7.3 and 7.4) the evaluation

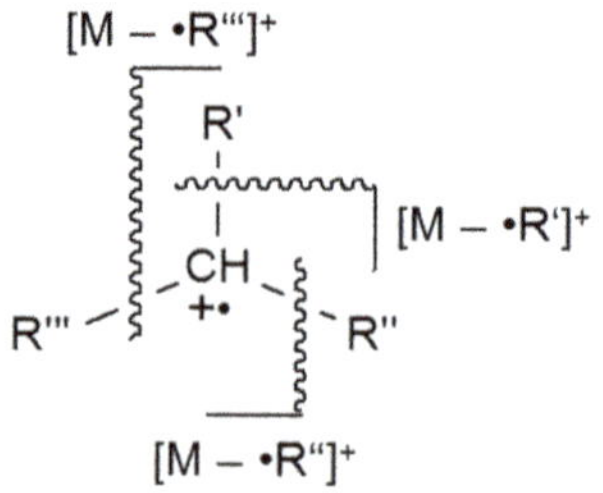

Scheme 7.5 Fragmentation of the weakest bond, that is, the one at a highly substituted carbon, leads to three complementary fragment ions and the
loss of three different radical species •R′, •R″ and •R‴ in this case the
rest are different.

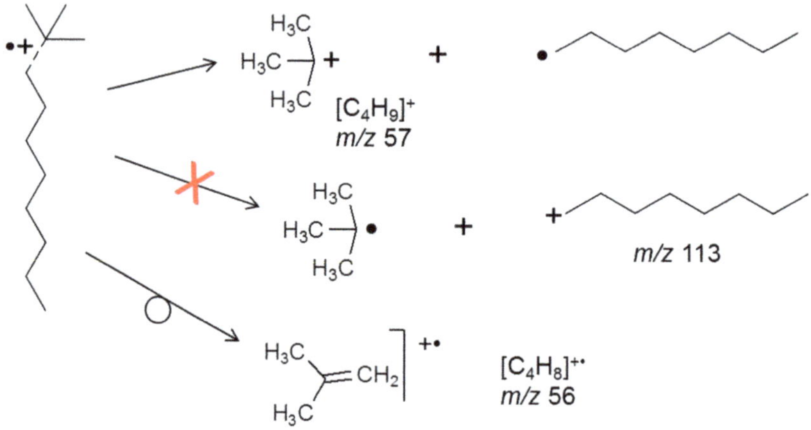

Scheme 7.6 The molecular ion of 2,2-dimethyldecane delivers the thermodynamically most stable fragment ions by cleavage of the weakest bond.
Here, the tertiary $[(CH_3)_3C]^+$ ion (a $[C_nH_{2n+1}]^+$ type ion) at m/z 57 is
formed because the IE of ($•C_4H_9$) is smaller than the IE of ($•C_8H_{17}$).
The competing primary n-alkyl ion at m/z 113 is barely observed in
the EI mass spectrum shown in Figure 7.5. The positive inductive stabilisation of the three methyl substituents explains the low IE of the
tertiary $•C_4H_9$ radical. The IE is much lower than that of the primary
octyl radical $•C_8H_{17}$, which is decisive for the competition for the
charge and also for the whereabouts of the unpaired electron (see
below).[1–3] Additionally, the radical cation of 2-methyl-propene, that is
an $[C_nH_{2n}]^{+•}$ ion, at m/z 56 is formed by the loss of octane upon a
hydrogen shift rearrangement reaction.

of competing fragmentation reactions with an analogous fragmentation mechanism is possible (Scheme 7.7). The thermochemistry gives
a good rationale for this issue: In simple bond cleavage reactions the
fragment with the lower IE retains the charge and the fragment with
the higher IE will be expelled as the radical (compare Scheme 7.6 and
Figure 7.13).

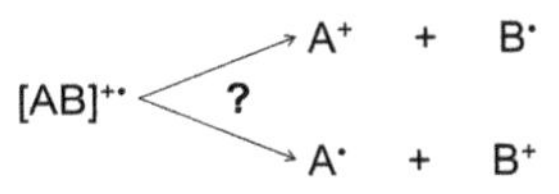

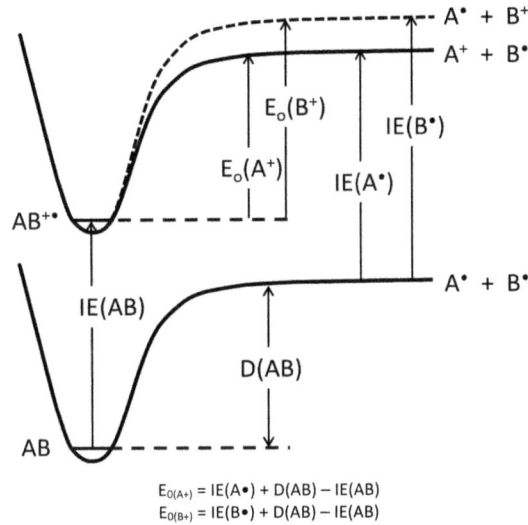

Scheme 7.7 Competing fragmentation reactions for the simple bond cleavage of a hypothetical biatomic molecular ion [AB]⁺•. The general reaction follows the scheme $OE^{+\bullet} \rightarrow EE^{+} + R^{\bullet}$ and proceeds *via* a similar loose transition state.

$$E_{0(A+)} = IE(A^{\bullet}) + D(AB) - IE(AB)$$
$$E_{0(B+)} = IE(B^{\bullet}) + D(AB) - IE(AB)$$

Figure 7.13 Vertical transition from a linear bi-atomic molecule AB to the respective molecular ion [AB]⁺• and the competing fragmentation reactions to the product ions A⁺ or B⁺. The energy balance is simplified by the assumption that the complementary product ions are formed *via* a loose transition state without a reverse energy barrier (Figure 5.12 in Chapter 5).[2,7,8]

This assumption relies on the stability of the complementary product ions of fragmentation reactions without a reverse activation barrier, meaning that the loose transition state (TS) of the reaction lies late on the reaction coordinate and the energy of the respective TS is very similar to the products of the cleavage reaction, as described by the *Hammond Principle* (see Figure 5.12 in Chapter 5, and Figure 7.13).[2,7,8]

$$E_{0(A+)} = IE(A^{\bullet}) + D(AB) - IE(AB)$$
$$E_{0(B+)} = IE(B^{\bullet}) + D(AB) - IE(AB)$$

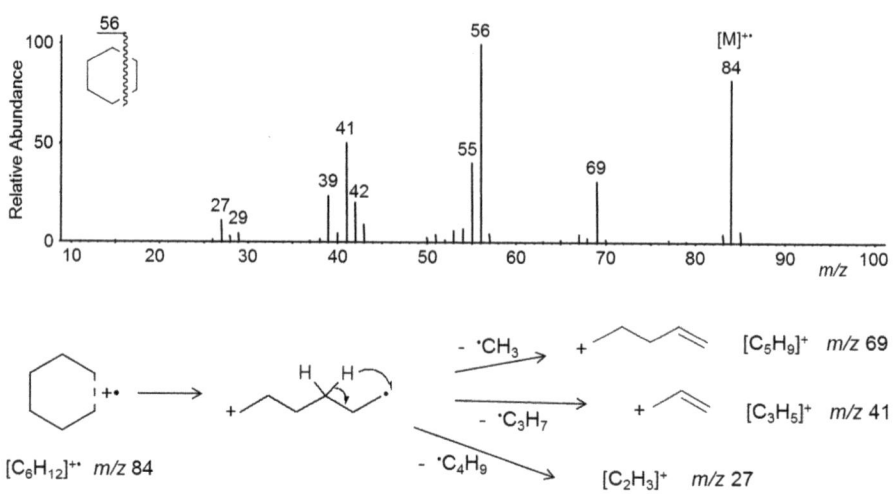

Figure 7.14 EI mass spectrum of cyclohexane. The unexpected loss of a methyl radical (15 u) from the cyclohexane molecular ion at *m/z* 84 is the result of a two-step fragmentation reaction including a hydrogen shift rearrangement. A series of alkene EE$^+$ cations at *m/z* 27, 41 and 69 with a Δm of 14 u (CH$_2$) is formed upon the loss of the respective alkyl radical R$^•$. Additionally, the elimination of ethylene (28 u) leads to the fragment ion [C$_4$H$_8$]$^{+•}$ at *m/z* 56, which retains the OE$^{+•}$ character of the molecular ion [M]$^{+•}$ owing to the loss of the neutral C$_2$H$_4$ molecule. The ion found at *m/z* 56 refers to [C$_4$H$_8$]$^{+•}$ and can either be cyclic, that is, the molecular ion cyclobutane, or linear. Spectra taken from the NIST Mass Spectral Library, Version 2.0f, 2008. Reproduced with permission.

7.2.5 Fragmentation of Activated Sigma Bonds: Allylic and Benzylic Cleavage

After the discussion of the fragmentation reactions in alkanes we now look at organic compounds with double bonds, for example alkenes, olefins and aromatics. Isolated double bonds in alkenes, but even more so the conjugated double bonds in aromatic ring systems, stabilise the EE$^+$ ions formed by loss of a radical through resonance (positive mesomeric effect + M). In alkenes, one π-electron out of the double bond is preferably lost in the ionisation process, as the sigma bond scaffold remains intact (Scheme 7.8). The α-bond relative to the position of the unpaired electron is subsequently cleaved leading to an allylic cation in the case of the alkenes. In Figure 7.15 an EI mass spectrum of an example alkene is shown and the fragmentation reactions which explain the generation of the ions in the spectrum are presented in Scheme 7.9.

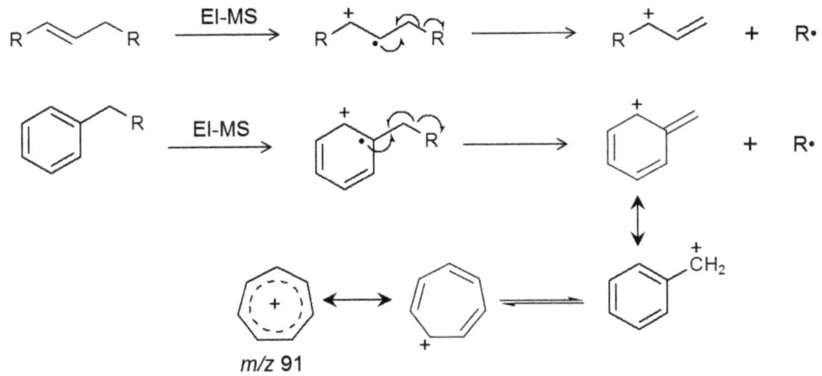

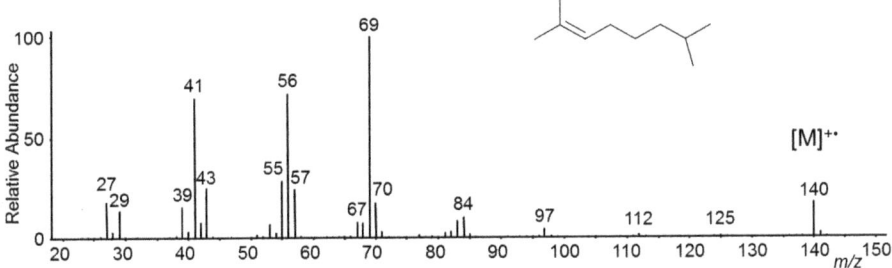

Scheme 7.8 Allylic and benzylic fragmentation pathway. Note that the allylic ion is resonance stabilised. In the presence of the alkylic substituents R, the molecular ion of an alkene can isomerise *via* hydrogen shift reactions hampering the localisation of the double bond. Additionally, the molecular ion allows rotation, making the identification of *cis/trans* and *E/Z* diastereomers *via* EI-MS impossible. The benzylic cation can further rearrange in a ring extension reaction to form the 7-membered ring system, that is the aromatic tropyllium cation $[C_7H_7]^+$ at *m/z* 91, a typical fragment ion of the substituted benzenes (compare Figures 7.18, 7.19 and 7.20).

Figure 7.15 EI mass spectrum of 2,7-dimethyl-2-octene. Spectrum taken from the NIST Mass Spectral Library, Version 2.0f, 2008. Reproduced with permission.

Halogenated hydrocarbons show typical fragmentation patterns in EI-MS as documented in Figures 7.16 and 7.18. The signal pattern found for the molecular ion $[M]^{+\bullet}$ of 1-chloro-3-methyl-butene (at *m/z* 104 and 106) reflects the natural abundance of the chlorine isotopes ^{35}Cl and ^{37}Cl as discussed in Section 5.2.3 in Chapter 5. The 1-chloro-3-methyl-butene is either ionised at the chlorine substituent, which has three unbound electron pairs available, or at the double bond with its π-electrons as depicted in Figure 7.16. The allylic

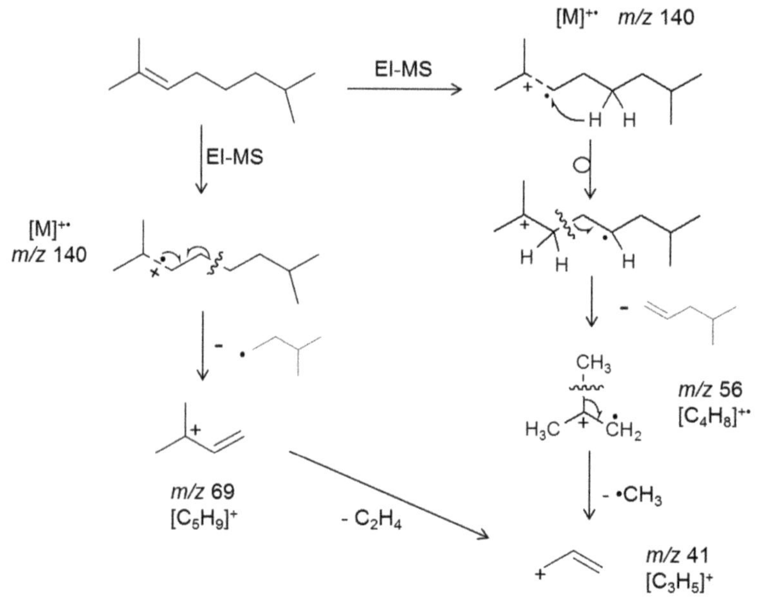

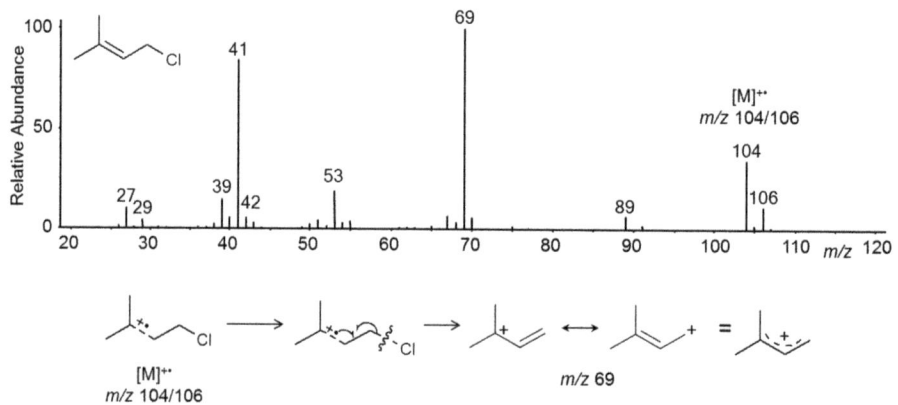

Scheme 7.9 Fragmentation pathways explaining the formation of the prominent fragment ions at m/z 69, 56 and 41 in the EI mass spectrum of 2,7-dimethyl-2-octene (Figure 7.15).

Figure 7.16 EI mass spectrum of 1-chloro-3-methyl-butene. The molecular ion $[M]^{+\bullet}$ at m/z 104 relates to the monoisotopic composition $[^{12}C_5{}^1H_9{}^{35}Cl]^{+\bullet}$ and the ion at m/z 106 to $[^{12}C_5{}^1H_9{}^{37}Cl]^{+\bullet}$, respectively. Spectrum taken from the NIST Mass Spectral Library, Version 2.0f, 2008. Reproduced with permission.

cleavage enforces the loss of the chlorine atom and forms the resonance stabilised allylic ion at m/z 69 (see fragmentation mechanism inset in Figure 7.16). Finally, the ion at m/z 41 results from subsequent

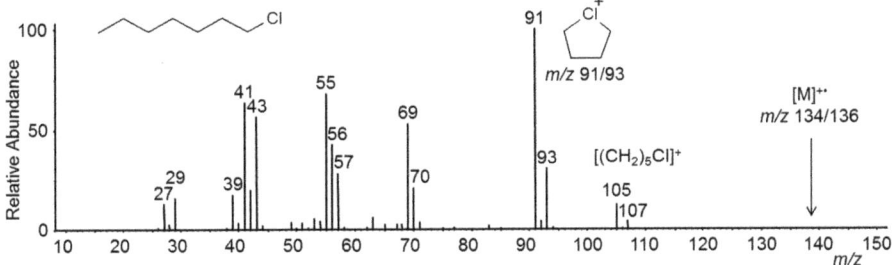

Figure 7.17 EI mass spectrum of 1-chloro-heptane. The identity of the ion at *m/z* 91 $[C_4H_8{}^{35}Cl]^+$ can be clarified and reliably assigned by accurate ion mass measurement and the characteristic isotopic pattern of chlorine, which shows the presence of the characteristic signal of the ion with the monoisotopic composition of $[{}^{12}C_4{}^1H_8{}^{37}Cl]^+$ at *m/z* 93 with an abundance of about a third of the abundance of the ion at *m/z* 91. Spectrum taken from the NIST Mass Spectral Library, Version 2.0f, 2008. Reproduced with permission.

ethylene ($C_2H_4 = 28$ u) neutral loss, as shown in Scheme 7.9. Both ions do not contain a chlorine substituent, as the respective satellite signals at *m/z* 71 and 43 are missing.

This criterion is important for the interpretation of the EI mass spectrum shown in Figure 7.17. Here, the ion at *m/z* 91 is the product of a special reaction of the ionised chloro-substituent with the δ-methylene carbon of the alkyl-chain leading to a five-membered ring structure (see inset). The presence of the halogen Cl is documented by the characteristic isotopic signal of $[C_4H_8{}^{37}Cl]^+$ at *m/z* 93. The formation of these ring-type fragment ions found at *m/z* 91 in the EI mass spectrum of 1-chloroheptane is a typical fragmentation of halogen-alkanes (X = Cl and Br) with sufficiently long chains. An independent option to further clarify the identity of the ion at *m/z* 91 would be the measurement of the accurate ion mass to confirm the composition to be $[C_4H_8{}^{35}Cl]^+$. On this basis alternative compositions such as $[C_7H_7]^+$ (also *m/z* 91) can be firmly excluded owing to the unique monoisotopic mass.

In the case of substituted aromatic compounds a benzylic fragment ion is formed as shown in Schemes 7.6 and 7.8 and in Figures 7.18 and 7.19. The α-bond relative to the double bond is *activated* for the subsequent fragmentation, as the ionisation of the double bond is clearly preferred. The resonance stabilisation of the EE^+ product ion after the radical loss process facilitates the allylic-, benzylic- (Figures 7.19 and 7.20) and also the α-cleavage of the amines for example (Scheme 7.10). The EI mass spectra of diastereomers, for example *Z/E* or *cis-* and *trans*-alkenes are typically very similar as the respective $[M]^{+\bullet}$

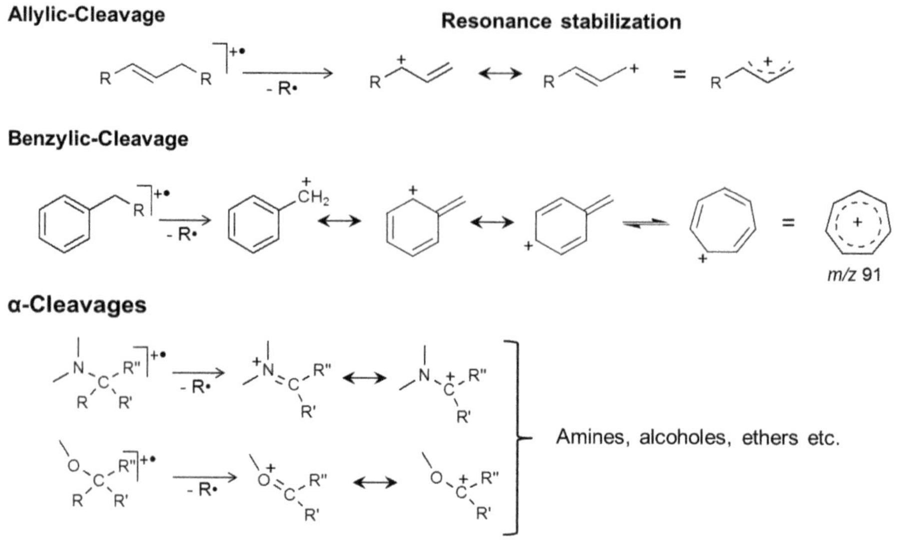

Allylic-Cleavage   **Resonance stabilization**

Benzylic-Cleavage

m/z 91

α-Cleavages

Amines, alcoholes, ethers etc.

Scheme 7.10   Fragmentation of activated sigma bonds: allylic-, benzylic- and α-cleavages of amines, alcohols and ethers.

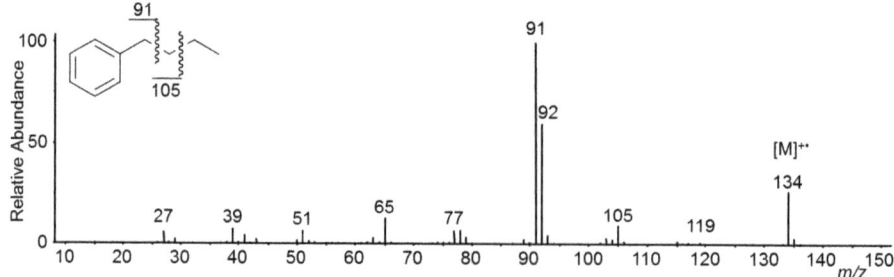

Figure 7.18   EI mass spectrum of butyl benzene with the prominent benzylic fragmentation reaction product $[C_7H_7]^+$ at *m/z* 91 and the product of the McLafferty rearrangement reaction at *m/z* 92 (see Section 7.2.9). The most prominent fragment ion formed by the loss of an alkyl radical is the tropyllium ion $[C_7H_7]^+$ at *m/z* 91. The signal intensity of similar fragment ions with increasing side chain lengths (see inset) are of lower abundance owing to the less effective stabilisation, compare these with the signal intensity of the ions at *m/z* 91 and *m/z* 105. Spectrum taken from the NIST Mass Spectral Library, Version 2.0f, 2008. Reproduced with permission.

molecular ions of the alkenes are highly excited and isomerise meaning that individual structural assignment is not usually possible *via* EI-MS.

Inspection of the EI mass spectra of the different constitutional isomers of butyl benzene shown in Figures 7.18 and 7.19 together with Schemes 7.8 and 7.10 highlights the potential of EI-MS to

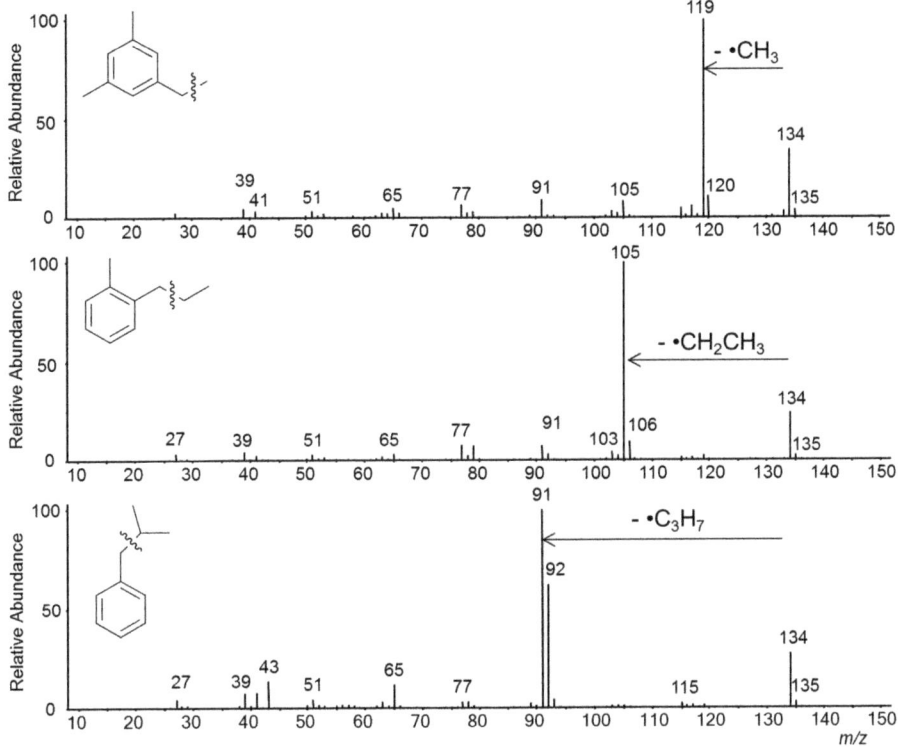

Figure 7.19 EI mass spectra of three constitutional alkyl benzene isomers. The benzylic-cation $[C_7H_7]^+$ at m/z 91 is very stable, indicating that the *benzylic* C^α-C^β-bond is therefore very weak and is preferably cleaved (see also Figure 7.20 and Scheme 7.10). The fragment ions of this process reflect the other remaining substituents at the respective aromatic precursor molecules, as no additional substituent other than the $[C_7H_7]^+$ at m/z 91 are formed. A methyl substituent at the benzylic fragment ion is responsible for a 14 u mass shift, $[C_8H_9]^+$ at m/z 105, and a dimethyl-analogue is found at $[C_9H_{11}]^+$ at m/z 119.[43] The formation of the ion at m/z 92 *via* the McLafferty rearrangement reaction will be discussed in Section 7.2.9. Spectra taken from the NIST Mass Spectral Library, Version 2.0f, 2008. Reproduced with permission.

distinguish analytes, especially to identify constitutional isomers, on the basis of characteristic fragmentation patterns. All four isomers have the same molecular mass (134 u), composition $C_{10}H_{14}$ and m/z ratio of the molecular ion $[M]^{+\bullet}$ (at m/z 134), but the differing aromatic substitution patterns leads to individual sets of fragment ions. The straight forward interpretation on the basis of the prominent benzylic fragmentation pathway for this class of analytes delivers a reliable rationale for the formation of the different fragment ions in the spectra and allows reliable structural assignment and analyte identification.

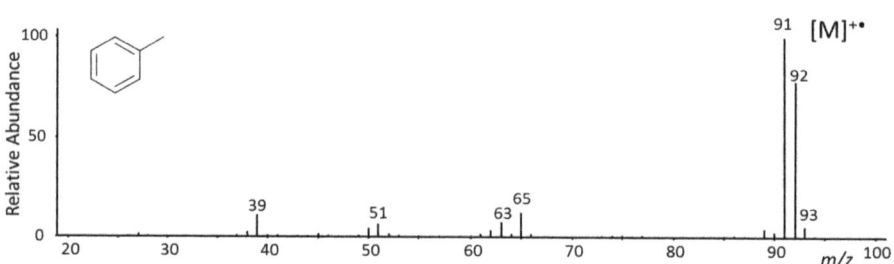

Figure 7.20 EI mass spectrum of toluene with the molecular ion [M]$^{+\bullet}$ at *m/z* 92. The strong signal corresponding to the loss of a hydrogen atom H$^\bullet$, that is, the [M − 1]$^+$ signal at *m/z* 91, is the product ion of the benzyl cleavage [C$_7$H$_7$]$^+$. The high stability of the [C$_7$H$_7$]$^+$ tropyllium cation shown in Scheme 7.8 compensates for the substantial energy demand of the typically unfavoured loss of the reactive and unstable hydrogen atom. Spectrum taken from the NIST Mass Spectral Library, Version 2.0f, 2008. Reproduced with permission.

7.2.6 Fragmentation of Activated Sigma Bonds: (Alpha) α-bond Cleavage Relative to Heteroatoms

Similar to the unsaturated hydrocarbons and substituted aromatic compounds that we have just discussed (see allylic and benzylic fragmentation reactions), the unbound, free electron pairs of hetero-atoms such as nitrogens, oxygens and halogens also allow facilitated ionisation in EI-MS. This applies for amines, alcohols, ethers, aldehydes, ketones, carboxylic acids and esters for example (Scheme 7.10) in which the sigma bond skeleton remains intact even though an electron was lost in the ionisation process. The molecular ions [M]$^{+\bullet}$ of these compounds exhibit a prominent fragmentation pathway, in which the α-bond dissociates homolytically. This is observed because the strong Lewis acidity of the unpaired electron, as well as the positive charge located at the heteroatom, polarizes the neighbouring sigma bonds. Additionally, a resonance stabilised EE$^+$ ion is formed after the loss of a radical as shown in Schemes 7.10 and 7.11.

In Figure 5.8 in Chapter 5 the EI mass spectrum of dodecanol is shown in which the product ion of the α-bond fragmentation at *m/z* 31 [CH$_2$=OH]$^+$ is observed. However, the molecular ion [M]$^{+\bullet}$ is not found in the spectrum, which is typical for long chain (higher) alcohols, hampering the identification and determination of the alkyl-chain length. Owing to the facilitated loss of H$_2$O the [M − H$_2$O]$^{+\bullet}$ product ion at *m/z* 168 is observed (M − 18 u) with the highest *m/z* value. In total the EI mass spectra of alcohols often resemble the ones of alkenes as the [M- H$_2$O]$^{+\bullet}$ ions formed primarily from

Atom positions

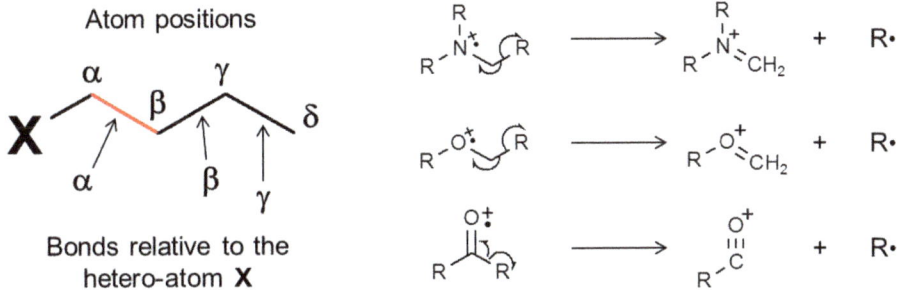

Bonds relative to the
hetero-atom **X**

Scheme 7.11 Notation of atom positions and sigma bonds. The (alpha) α-bond
fragmentation of an activated sigma bond (alpha bond red colour
coded) adjacent to the hetero-atoms **X** = **NH₂, OH, OR, SH**, is the
prominent cleavage reaction observed in the EI-MS of respective odd
electron molecular ions [M]⁺• of amines, alcohols, ethers, ketones,
aldehydes, carboxylic acids and esters.

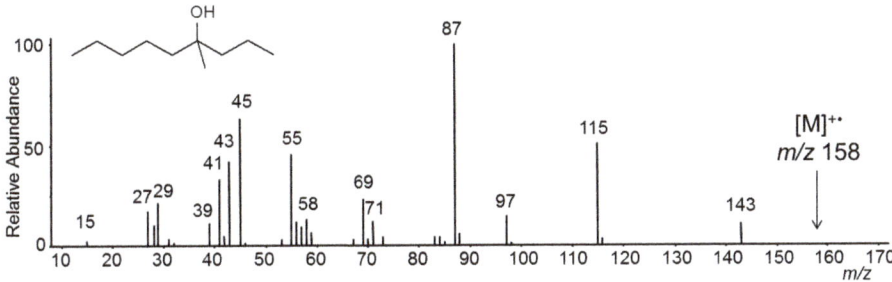

Figure 7.21 The molecular ion [M]⁺• at *m/z* 158 of 4-methyl-4-nonanol is not
observed in the EI mass spectrum. The cleavage of the C–C bond in
the α-position of the hydroxy-moiety is frequently found in the EI-MS
of alkanols because stable oxonium ions are formed. Spectrum taken
from the NIST Mass Spectral Library, Version 2.0f, 2008. Reproduced
with permission.

the alcohol molecular ions correspond to the radical cations of the
respective alkenes. To clarify the structure of an alcohol analytes
derivatisation and EI-MS of the respective tri-methyl-silyl-ether
(TMS) can be helpful (we can compare the water loss rearrangement
reactions as discussed in Section 7.2.8).

In the spectrum shown in Figure 7.21 three complementary
fragment ions are formed by three α-cleavage reactions of the
4-methyl-4-nonanol molecular ion [M]⁺•. These ions are found at *m/z*
143 (loss of •CH₃), *m/z* 115 (loss of •CH₂CH₂CH₃) and at *m/z* 87 (loss of
•C₅H₁₁). In contrast to the considerations on the stability of the EE⁺
ions formed in competing radical loss reactions (see Section 7.2.4),
here the stability of the expelled radicals governs the extent of product

ion formation. The stability ordering of the radicals is: $^{\bullet}CH_3 < {}^{\bullet}C_3H_7 < {}^{\bullet}C_5H_{11}$. The abundance ordering of the $[R_1R_2C=OH]^+$ oxonium ions is found accordingly: m/z 87 > m/z 115 > m/z 142 as documented in Figure 7.21. This consideration is also valid for the three α-cleavage reactions found in the EI mass spectrum of the tertiary amine shown in Figure 7.23.

Often the molecular ions of alcohols are missing in the EI mass spectra (Figures 7.8 and 7.21). To overcome this problem and allow structural identification, for example the determination of the alkyl chain length, the formation of an appropriate derivative is recommended (*e.g.* R–OH + $(CH_3)_3$SiCl → R–O–Si$(CH_3)_3$ + HCl). The EI mass spectrum of the respective TMS derivatives documents predictable fragmentation reactions (formation of the $[M - {}^{\bullet}CH_3]^+$ ion) and characteristic fragment ions as shown in Figure 7.22 and Scheme 7.12.

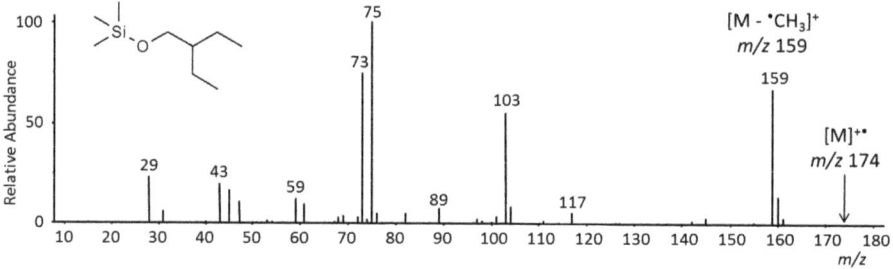

Figure 7.22 EI mass spectrum of the TMS-derivative of 2-ethyl-1-butanol with the predicted $[M - {}^{\bullet}CH_3]^+$ ion at m/z 159. Derivatisation is an important part of sample preparation in GC-MS applications. Derivatisation also serves well to convert a less volatile (polar) compound to a (less polar) volatile derivative, which can be more easily transferred to the gaseous phase by heating (*e.g.* TMS-derivatisation of alcohols).

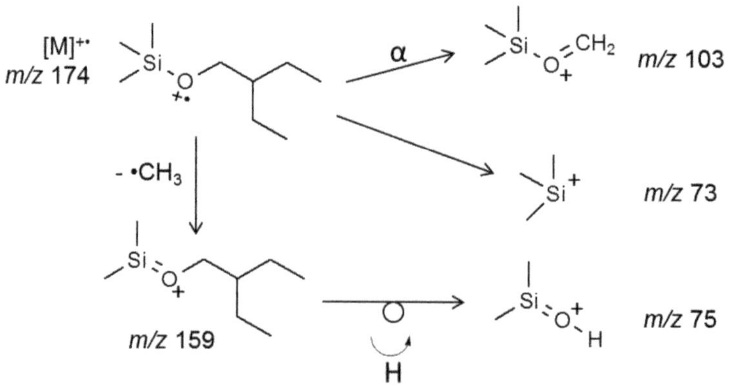

Scheme 7.12 Fragmentation pathways of the molecular ion of 3-ethylbutoxy-trimethylsilane at m/z 174.

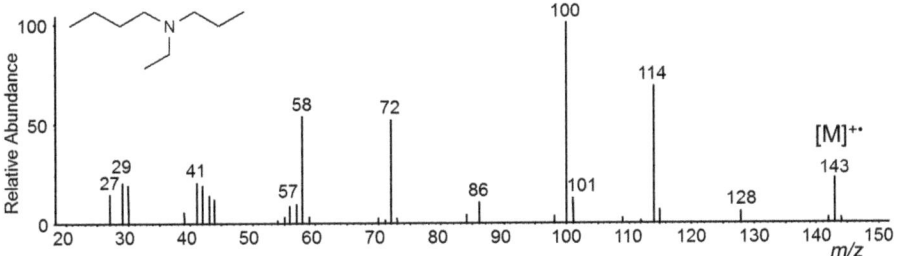

Figure 7.23 EI mass spectrum of *N*-ethyl-*N*-propyl-butanamine ($C_9H_{21}N$). The α-cleavage of the molecular ion of the tertiary amine *N*-ethyl-*N*-propyl-butanamine at *m/z* 143 (uneven molecular mass, compare with the nitrogen rule; see Section 7.1.2) delivers three complementary ammonium fragment ions $[R_1R_2N=CH_2]^+$ at *m/z* 128, *m/z* 114 and *m/z* 100 (compare with Schemes 7.10 and 7.11). Spectrum taken from the NIST Mass Spectral Library, Version 2.0f, 2008. Reproduced with permission.

The EI mass spectrum of the *N*-ethyl-*N*-propyl-butanamine shows the α-cleavage reactions as highlighted above. Additionally, three significant fragment ions are observed at *m/z* 58, 72 and 86 in Figure 7.23. The secondary fragmentation reaction starting from the ammonium ions at *m/z* 100 delivers the ion at *m/z* 58 by loss of propene C_3H_6 as illustrated in Scheme 7.13. This reaction is called the "*onium*" reaction because it is also relevant for amm*onium*, ox*onium*, sulph*onium*, and phosph*onium* ions (see also Scheme 7.14). The "*onium*" reaction is a typical fragmentation reaction of EE^+ ions with sufficiently long alkyl chain moieties in EI-MS allowing further loss of olefins: $EE^+ \rightarrow EE^+ +$ olefins (compare with Figure 7.14). Finally, the loss of an ethylene molecule C_2H_4 can be explained by a rearrangement reaction proceeding *via* a 6-membered transition state, that is the so called McLafferty reaction (analogous to the thermal ester pyrolysis reaction; see Schemes 7.13 and 7.24 and Section 7.2.4).[1]

In case we find more than one hetero-atom in the analyte structure, several starting structures are considered to explain the EI-MS fragmentation pathways of *N,N*-dimethyl-benzamide as Figure 7.24 and Scheme 7.15 show. In particular, the significant formation of the $[M - H^·]^+$ ion at *m/z* 148 by an α-cleavage reaction (loss of a hydrogen atom) should be noted. The nitrogen rule (a compound with an uneven number of nitrogens has an uneven mass of the molecular ion $[M]^{+·}$) as well as a close evaluation of the signal intensities of the ions at *m/z* 148 and *m/z* 149 in relation to the composition of *N,N*-dimethyl-benzamide leads to a correct assignment.

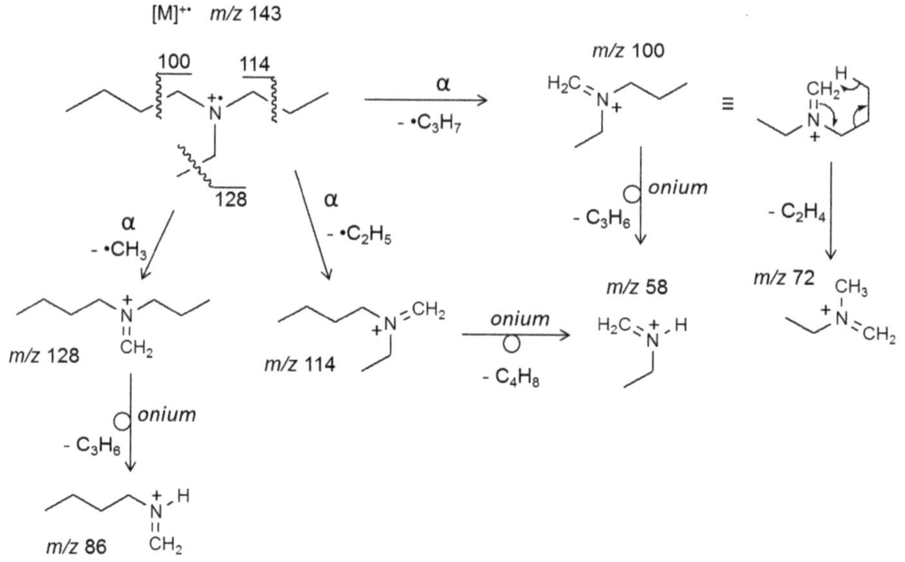

[M]⁺˙ *m/z* 143

Scheme 7.13 Fragmentation pathways of the molecular ion of *N*-ethyl-*N*-methyl-butane-amine at *m/z* 143.

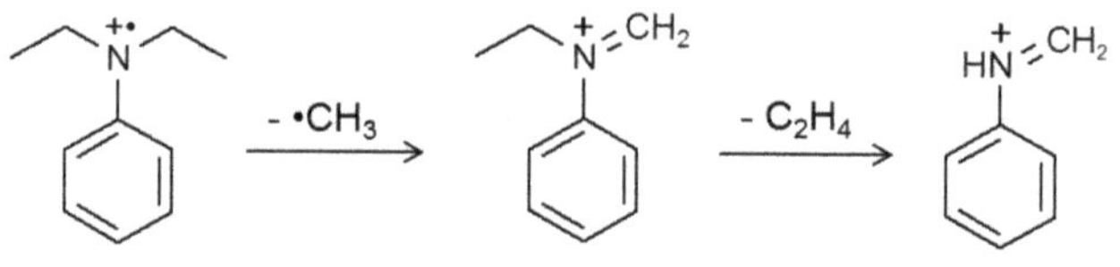

Scheme 7.14 Sequential fragmentation reactions of the molecular ion of *N,N*-diethyl-*aniline*. The initial α-cleavage reaction (loss of ˙CH₃) is followed by the *onium-reaction* of the EE⁺ ammonium ion formed in the primary step.

7.2.7 Isomerisation Reactions Prior to Fragmentation: Hydrogen Shift Rearrangement Reactions in Alicyclic Compounds

As already shown in Figure 5.14 (Chapter 5) highly excited molecular ions [M]⁺˙ can isomerise in EI-MS before a monomolecular fragmentation reaction leads to the formation of product ions in subsequent steps. Evidence for that can be found in the EI mass spectrum of cyclohexanol, presented in Figure 7.25. The molecular ion [M]⁺˙ of cyclohexanol with both the unpaired electron and the positive charge at the oxygen triggers an initial α-cleavage, which leads to a distonic ion as Scheme 7.16 illustrates. The subsequent rearrangement reaction, that is a hydrogen shift isomerisation gives

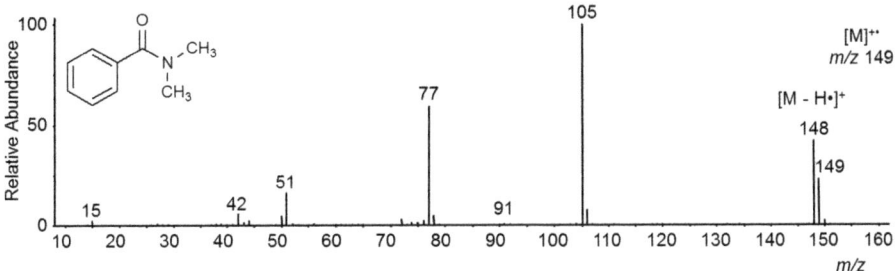

Figure 7.24 EI mass spectrum of *N,N*-dimethyl-benzamide. The [M-H•]+ at *m/z* 148 is formed by an α-cleavage as Scheme 7.15 illustrates. Spectrum taken from the NIST Mass Spectral Library, Version 2.0f, 2008. Reproduced with permission.

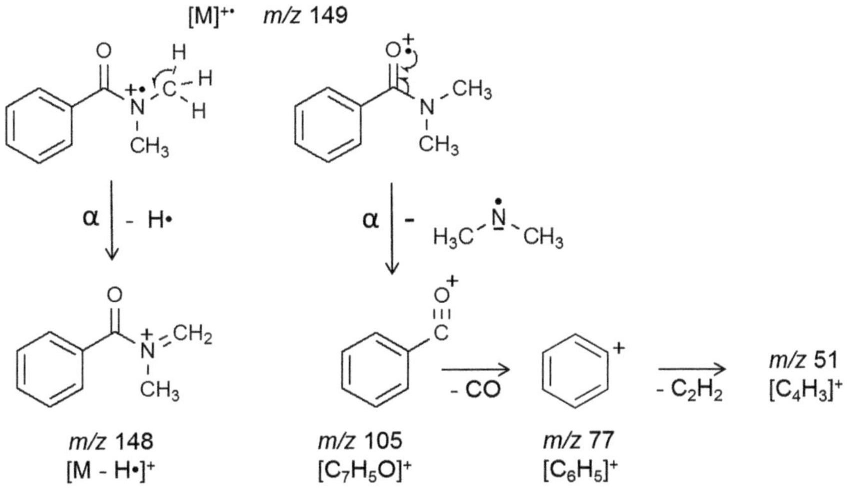

Scheme 7.15 Fragmentation pathways of the molecular ion of *N,N*-dimethyl-benzamide at *m/z* 149 (compare with Figure 7.22). Once the EE+ ions are formed by the loss of a radical from the OE+• molecular ion, sequential fragmentation reactions will also form EE+ + neutral molecules (here: the loss of CO and an alkyne, such as C₂H₂). The formation of two open shell species (OE+• + •R) from a closed shell EE+ precursor ion is very unlikely for energetic reasons (Figure 7.14).

rise to a stabilised ion (structure **b** in Scheme 7.16), as the primary radical site (structure **a** in Scheme 7.16) of the linear molecular ion acts as an acceptor site for the hydrogen atom migration. The final step is again an α-cleavage reaction relative to the radical site which leads to the loss of the ethyl radical (a fragment that is not available in cyclohexanol) and the formation of the significant ion at *m/z* 57 (Figure 7.25). If an alkyl substituent is present in position 2 of the

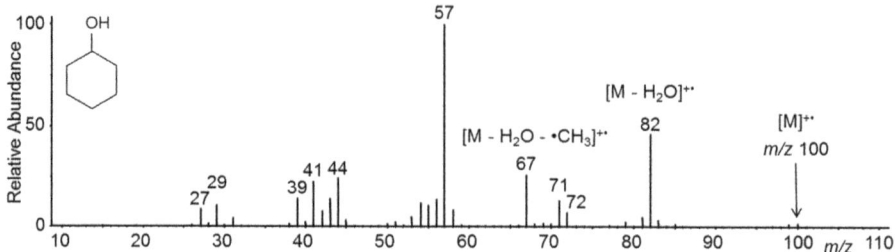

Figure 7.25 EI mass spectrum of cyclohexanol. The formation of the prominent fragment ion at *m/z* 57 is explained in Scheme 7.17. Spectrum taken from the NIST Mass Spectral Library, Version 2.0f, 2008. Reproduced with permission.

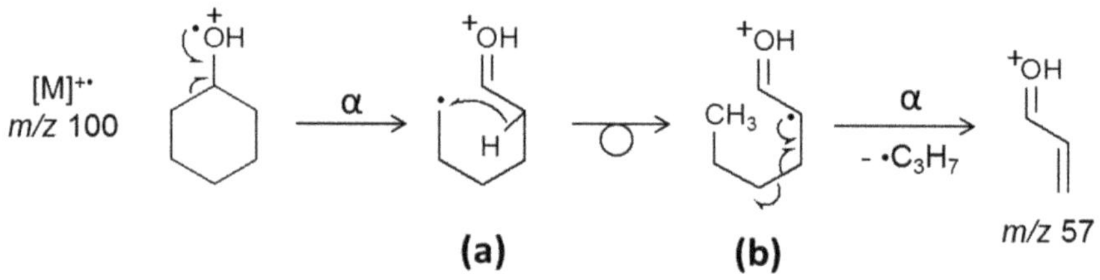

Scheme 7.16 Fragmentation pathway of the molecular ion of cyclohexanol at *m/z* 100. The initial α-cleavage delivers a *distonic* ion, in which the unpaired electron and the charge are not located at the same site in the ion. This ion performs a hydrogen shift isomerisation that delivers a resonance stabilised secondary radical (**b**) that is substantially more stable than the primary radical (**a**). The terminal α-cleavage enables the loss of the ethyl radical and the formation of the most abundant ion in the spectrum at *m/z* 57 (see Figure 7.25).

cyclohexyl-ring as illustrated in Scheme 7.17, the initial α-cleavage reaction is biased and the fragmentation leading to the mass shifted fragment is favoured owing to the higher stability of the distonic ion formed in the first place. An analogous consideration explains the important features of the EI-MS fragmentation behaviour of the molecular ion of 2,3-dimethyl-cyclohexylamine shown in Figure 7.26 and Scheme 7.18 (the fragment ion at *m/z* 56 is of higher abundance than that at *m/z* 84).[1,2]

7.2.8 How Do Rearrangement Reactions Leading to the Loss of Very Stable Neutral Molecules Such as H_2O, NH_3, HCN, C_2H_4, CO, CO_2 and So Forth Occur?

We have already observed the elimination of small neutral molecules in the EI-MS fragmentation pathways of many organic compounds. The rearrangement reactions that take place prior to the respective

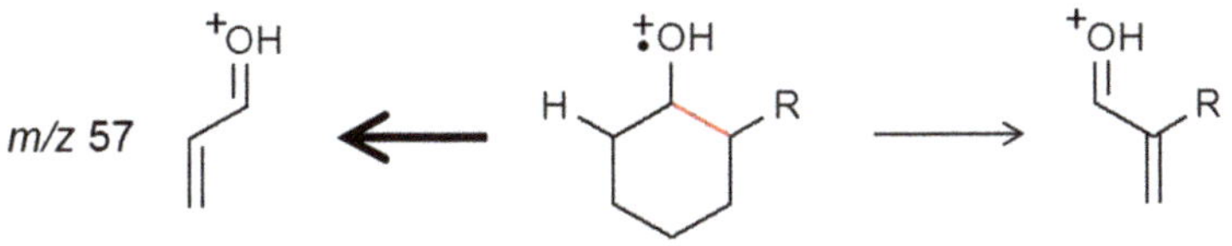

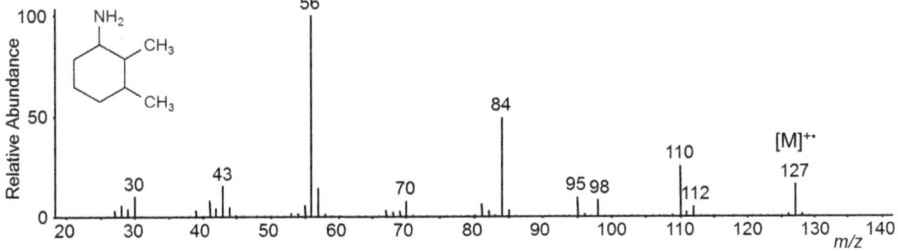

Scheme 7.17 The competing fragmentation pathways of the molecular ion of a substituted cyclohexanol. The depicted reaction pathways will eventually deliver two fragment ions mass shifted by the mass of the rest of R. The ion containing R will usually have a lower abundance than the ion at m/z 57, as the initial α-cleavage of the red colour coded α-bond will generate a slightly more stable *distonic* ion. In this case the secondary radical is inductively stabilised by the presence of the substituent R through hyper-conjugation (compare Schemes 7.16 and 7.18). The respective sequence of steps similar to those depicted in Scheme 7.15 will then lead preferably to the ion at m/z 57.

Figure 7.26 EI mass spectrum of 2,3-dimethyl-cyclohexylamine. The fragment signal at m/z 110 demonstrates the loss of ammonia (Δm 17 u = NH_3), a characteristic fragmentation of molecular ions of primary amines (here $[M]^{+\bullet}$ at m/z 127). Spectrum taken from the NIST Mass Spectral Library, Version 2.0f, 2008. Reproduced with permission.

loss of an olefine can be complicated, as seen in the case of alicyclic substituents (Scheme 7.18). However, molecules such as water, ammonia and carbon monoxide are expelled from open shell $[M]^{+\bullet}$ molecular ions (*e.g.* facile loss of water from aliphatic alcohols; see Figure 5.8, Chapter 5) as well as from EE^+ type fragment ions in subsequent processes.

Loss of small molecules:

(a) $OE^{+\bullet} \rightarrow OE^{+\bullet} +$ neutral molecule
(b) $EE^+ \rightarrow EE^+ +$ neutral molecule

In case (a) $OE^{+\bullet}$ type fragment ions are formed, which retain the unfavoured open shell character but the loss of small and very stable neutrals such as water obviously pays off energetically. Case (b) is the most significant pathway for the further fragmentation of EE^+ ions,

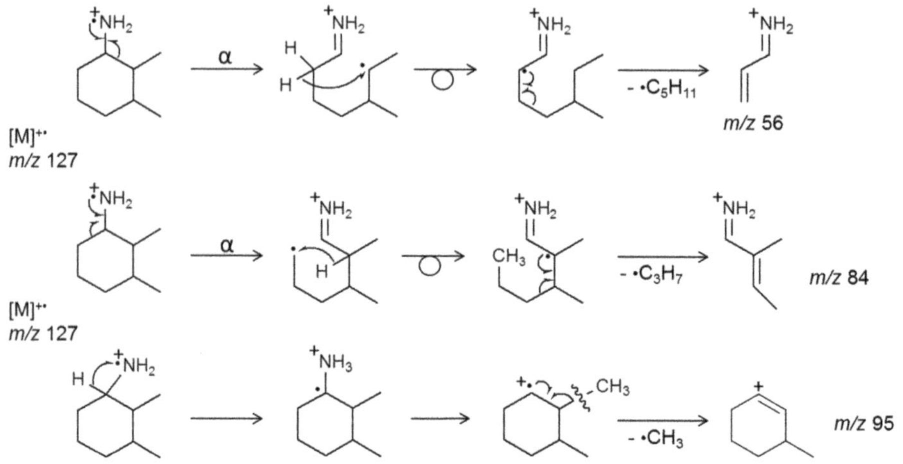

[M]⁺˙
m/z 127

[M]⁺˙
m/z 127

m/z 56

m/z 84

m/z 95

Scheme 7.18 The molecular ion of 2,3-dimethyl-cyclohexylamine exhibits analogous fragmentation pathways similar to those discussed in Schemes 7.16 and 7.17.

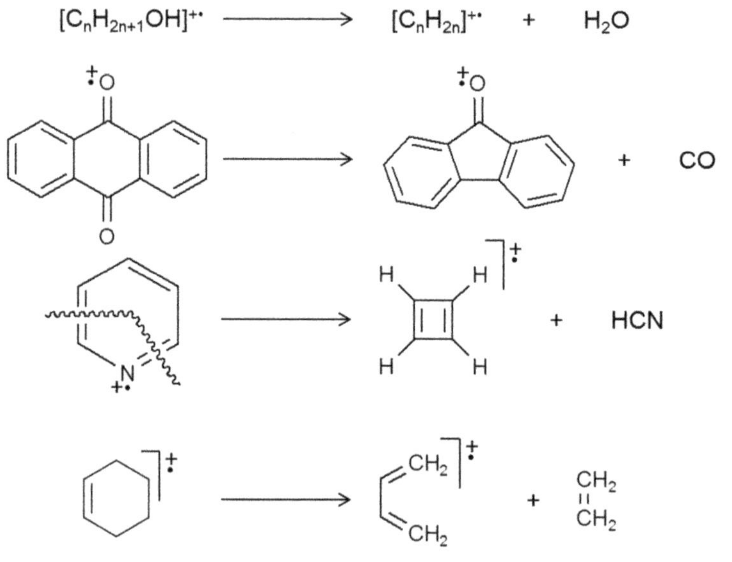

$$[C_nH_{2n+1}OH]^{+\cdot} \longrightarrow [C_nH_{2n}]^{+\cdot} + H_2O$$

+   CO

+   HCN

+ CH_2 $||$ CH_2

Scheme 7.19 Exemplary elimination reactions of neutral molecules from [M]⁺˙ molecular ions. The open shell character of the precursor ion is retained in the product ion: OE⁺˙ → OE⁺˙ + Neutral.

again for energetic reasons, as the loss of neutrals retain the closed shell electron configuration in the product ions formed (compare for example the *onium* reactions in Schemes 7.13 and 7.14 and the CO and C_2H_2 loss in Scheme 7.15).

Prominent examples for the loss of water H_2O (Δm 18 u) are found in the EI mass spectra of aliphatic alcohols (Scheme 7.19),

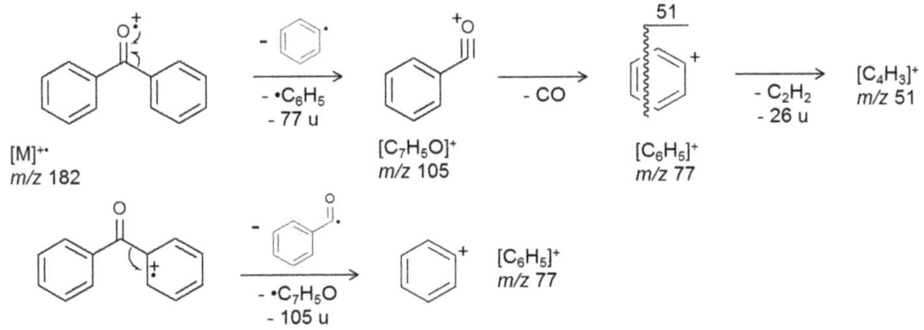

Scheme 7.20 The molecular ion of benzophenone exhibits a characteristic fragmentation pattern with the sequential loss of CO and C_2H_2 neutrals from the EE$^+$ ions. See also the EI-MS spectrum of benzophenone shown in Figure 7.4.

the EI-MS of cyclohexanol (Figure 7.25), 4-methyl-4-nonanol (Figure 7.19), and 1-dodecanol (Figure 5.8, Chapter 5). Ammonia loss NH_3 (Δm 17 u) is observed in the EI mass spectra of primary amine analytes, for example in the spectrum of dimethyl-cyclohexylamine (Figure 7.26).

The elimination of carbon monoxide (CO) (Δm 28 u) is found in many aromatic carbonyl compounds such as benzophenone and quinones (see Scheme 7.19 and the detailed fragmentation mechanism in Scheme 7.20). However, the mass difference of 28 u can also originate from the loss of ethylene C_2H_4 (as seen in *onium* and the McLafferty reaction pathways; *e.g.* Schemes 7.13 and 7.14). The loss of N_2 is rarely the reason for 28 Da neutral losses. A reliable and unambiguous assignment of the neutral that is actually lost in a fragmentation sequence is possible on the basis of an accurate mass measurement of the respective precursor and the product ion of that particular process, the mass difference of the two ions is thereby precisely determined and the neutral that was lost can be identified as the mass of CO, C_2H_4 and N_2 differs significantly. The mass of CO is 27.9949 u, the one of C_2H_4 is 28.0313 u and that of N_2 is 28.0061 u! Interestingly, CO loss is also the most prominent fragmentation reaction of the phenols (Scheme 7.21) and quinones (*e.g.* EI-MS of anthraquinone, see Scheme 7.19). Similar to the loss of CO in the phenols, hydrogen cyanide (HCN) (Δm 27 u) is eliminated from anilines as Scheme 7.22 illustrates. The characteristic loss of HCN (Δm 27 u) can also occur in pyridines as Scheme 7.19 highlights.

A mass difference of Δm 30 u is found less often, and can indicate the loss of formaldehyde, which is a characteristic elimination reaction found in the EI-MS of anisols as shown in Scheme 7.23.

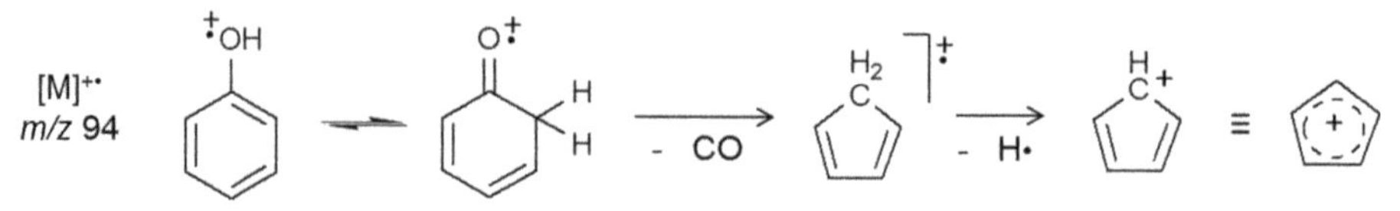

Scheme 7.21   The usually abundant molecular ion of phenol *m/z* 94 (and similar for the substituted derivatives) isomerises to a formal keto-tautomer that is able to expel the neutral CO molecule (Δm 28 u) in an $OE^{+\bullet}$ → $OE^{+\bullet}$ + neutral process. The subsequent loss of a hydrogen atom from $[C_5H_6]^{+\bullet}$ at *m/z* 66 leads to the $[C_5H_5]^+$ fragment ion, that is an EE^+ type ion at *m/z* 65.

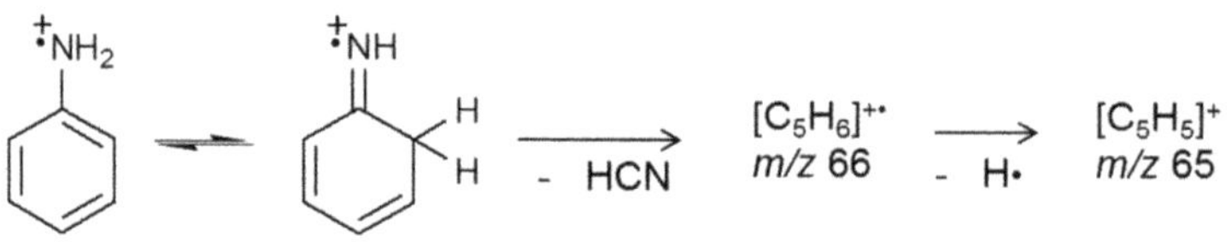

Scheme 7.22   Elimination of hydrogen cyanide (HCN) (Δm 27u) from aniline.

7.2.9   What Is the McLafferty Rearrangement Reaction?

Besides the reactions discussed above another important process involving a γ-hydrogen shift and β-bond cleavage, that is commonly called the *McLafferty* rearrangement reaction, is now discussed (Scheme 7.24). The prerequisite for the observed double bond shift and the cleavage, is a flexible chain of at least three sigma bonds that can adopt a 6-membered transition state and a hydrogen in the γ-position that can migrate as shown in Scheme 7.24. The reaction can also be formulated for the rearrangement reactions of the EE^+ precursor ions as indicated in Scheme 7.13 for an ammonium ion.

The loss of alkenes from carbonyl compounds is one prominent example of the *McLafferty* reaction as Figure 7.27 shows. In the EI mass spectrum of 4-octanone three fragment ions at *m/z* 86, *m/z* 99 and at *m/z* 58 are formed by the McLafferty reaction as illustrated in Schemes 7.25 and 7.26. In addition to this rearrangement reaction the α-cleavage and the subsequent loss of CO explains the rest of the major fragment ion signals in the EI mass spectrum of 4-octanone shown in Figure 7.27.

The *McLafferty* reaction is also relevant in the EI mass spectra of aromatic hydrocarbons with alkyl substituents of sufficient length. The EI mass spectrum of butyl-benzene shown in Figure 7.18 features a very important signal at *m/z* 92 that belongs to the product ion of the *McLafferty* reaction as shown in Scheme 7.27.

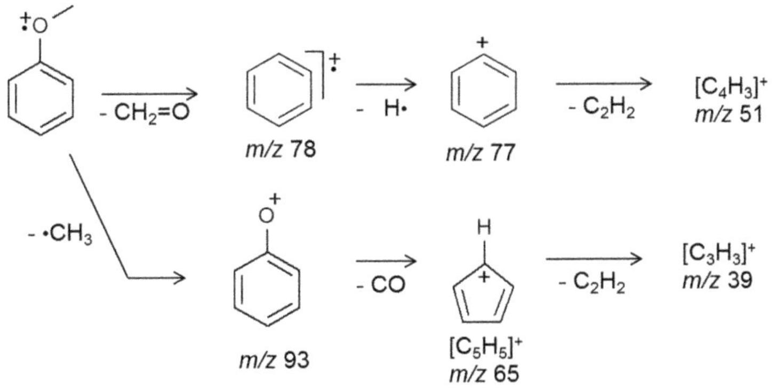

Scheme 7.23 In the EI mass spectrum of anisole the mass difference, Δm 30 u, between the molecular ion at m/z 108 and the fragment ion at m/z 78 results from the loss of formaldehyde ($CH_2=O$).

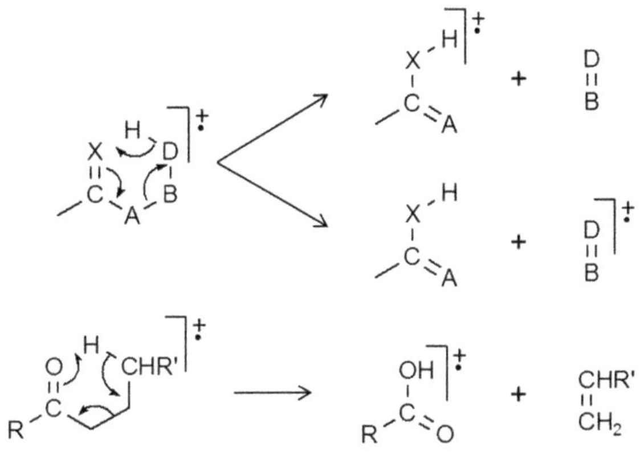

Scheme 7.24 Six-membered transition state of a *McLafferty* rearrangement reaction in general terms and in a molecular ion of an aliphatic ester. The reaction is not restricted to hydrocarbons but is found in a large variety of compounds, such as aldehydes, ketones and esters in which X can be C, O; A,B: CH_2, O; and D: CHR, CH_2. As indicated in the upper panel the reaction can lead to the presence of two complementary fragment ions in the EI mass spectrum, as the charge and the unpaired electron can reside at either of the two product species. The ionisation energy of the two potential product ion species is relevant for this competition (see also details of the *retro-Diels Alder* reaction).[9]

7.2.9.1 Ortho-*elimination or* Ortho-*effect in Aromatic Hydrocarbon Compounds*

An important elimination reaction for molecular ions of ortho-disubstituted aromatic hydrocarbon compounds relies on the close vincinity of the two substituents that interact as described

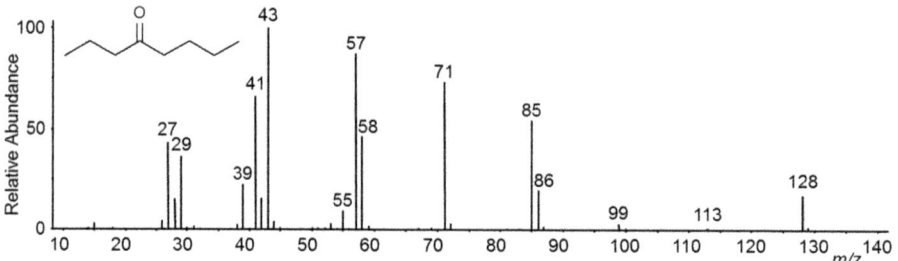

Figure 7.27 EI mass spectrum of 4-octanone. The fragment ion at *m/z* 86 is a product of the *McLafferty* rearrangement reaction (compare Schemes 7.24 and 7.25). Product ions of the two α-cleavage reactions deliver the ions at *m/z* 71 and *m/z* 85 and the subsequent CO loss reactions from these two primary fragment ions then explain the ions at *m/z* 43 and *m/z* 57 (compare with Schemes 7.11 and 7.15). Spectrum taken from the NIST Mass Spectral Library, Version 2.0f, 2008. Reproduced with permission.

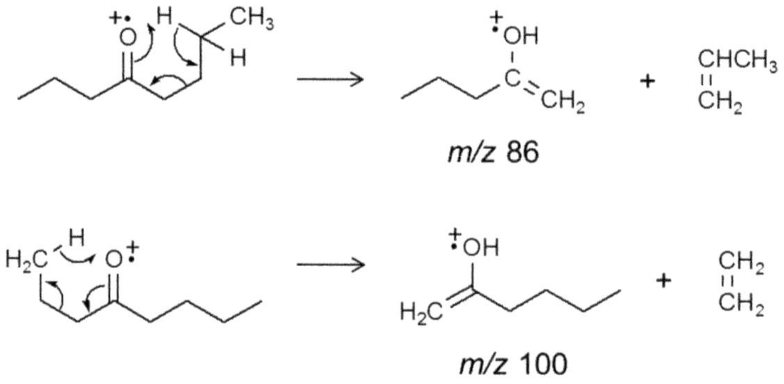

m/z 86

m/z 100

Scheme 7.25 *McLafferty* rearrangement reaction of the molecular ion of 4-octanone explaining the two complementary fragment ions at *m/z* 86 and at *m/z* 100 in the EI mass spectrum of 4-octanone (compare with Figure 7.27).

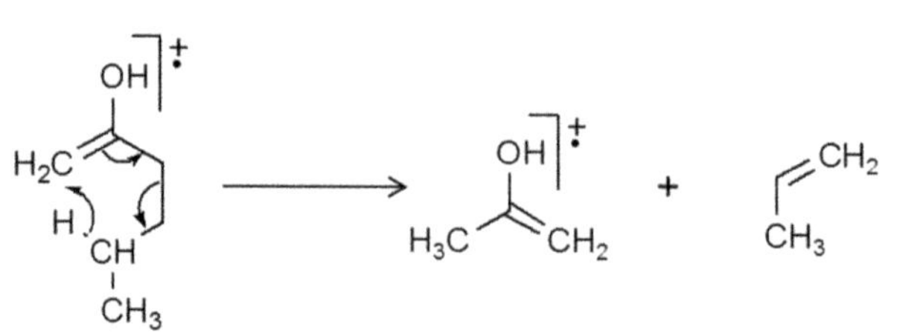

Scheme 7.26 The primary fragment ion at *m/z* 100 undergoes another *McLafferty* rearrangement reaction which gives rise to the prominent ion at *m/z* 58 and explains in turn the minimal abundance of *m/z* 100. Consecutive fragmentation reactions are discussed in Figure 5.14, in Chapter 5.

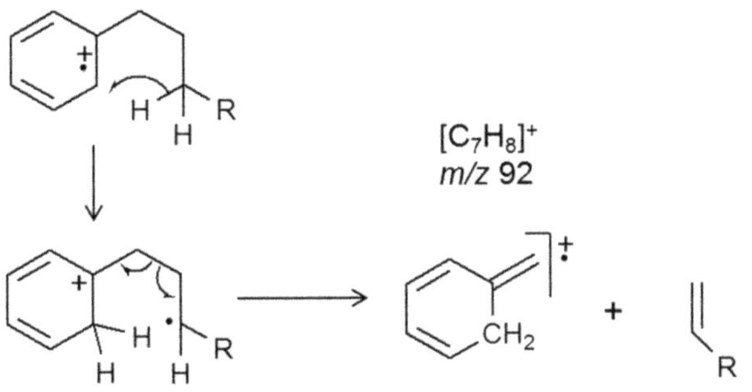

Scheme 7.27 *McLafferty* fragmentation reaction (McL) generating the product ion at *m/z* 92 found in the EI mass spectra of the aromatic hydrocarbons. The ion at *m/z* 92 originates from the loss of an alkene and the migration of a hydrogen atom in either a concerted or a stepwise mechanism (compare this with the EI mass spectrum of butyl benzene shown in Figure 7.18).[10]

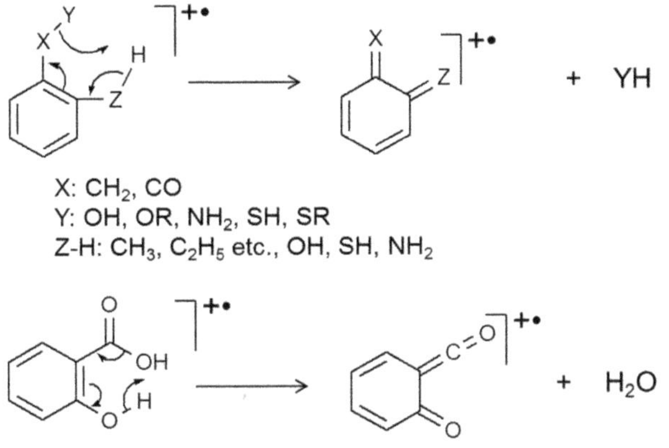

X: CH$_2$, CO
Y: OH, OR, NH$_2$, SH, SR
Z-H: CH$_3$, C$_2$H$_5$ etc., OH, SH, NH$_2$

Scheme 7.28 *Ortho*-elimination of a neutral molecule in the molecular ions of *ortho*-disubstituted aromatic compounds, for example the fragmentation of the molecular ion [M]$^{+\bullet}$ of salicylic acid (see lower panel).

in Scheme 7.28. A hydrogen transfer *via* a 6-membered transition state enables the loss of neutral molecules such as alcohols, water, ammonia and so forth. This reaction can be very helpful for the selective identification of constitutional isomers with different substitution patterns as shown in the comparison of the two isobaric (molecules of the same mass) esters presented in Figures 7.28 and 7.29 and Schemes 7.29 and 7.30, respectively.

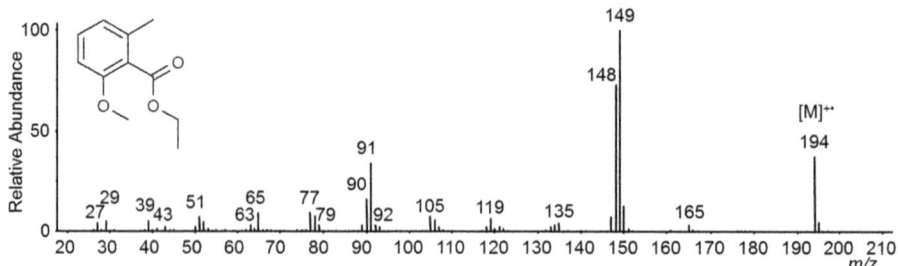

Figure 7.28 EI mass spectrum of 2-methoxy-6-methyl-benzoic acid ethyl ester. Spectrum taken from the NIST Mass Spectral Library, Version 2.0f, 2008. Reproduced with permission.

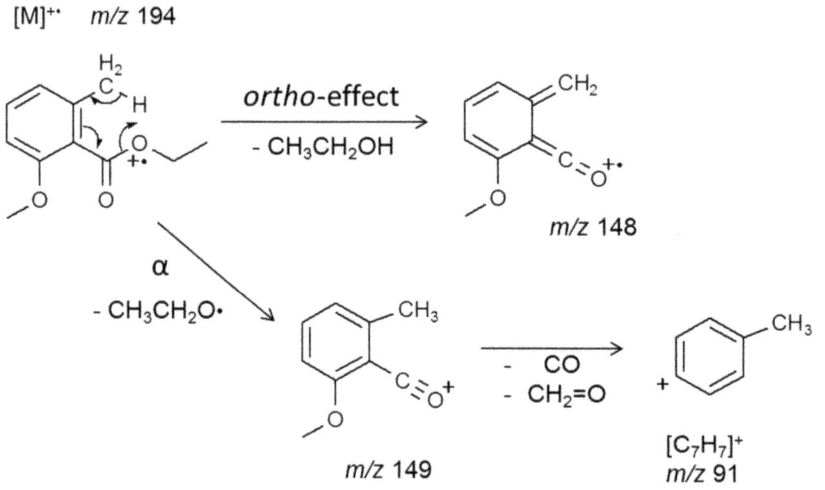

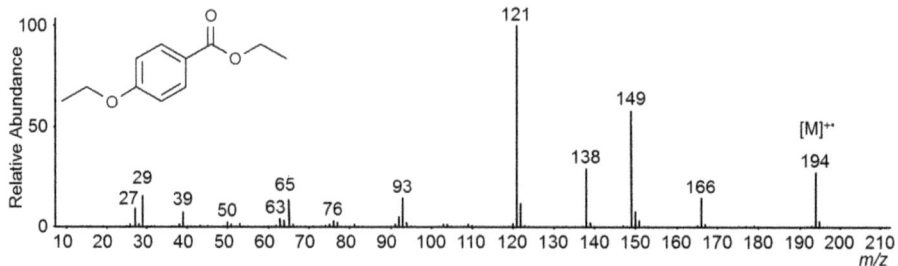

Scheme 7.29 Fragmentation reactions including the ortho-elimination of ethanol (Δm 46 u) only possible for the ortho-disubstituted 2-methoxy-6-methyl-benzoic acid ethylester.

Figure 7.29 EI mass spectrum of 4-ethoxy-benzoic acid ethylester. Spectrum taken from the NIST Mass Spectral Library, Version 2.0f, 2008. Reproduced with permission.

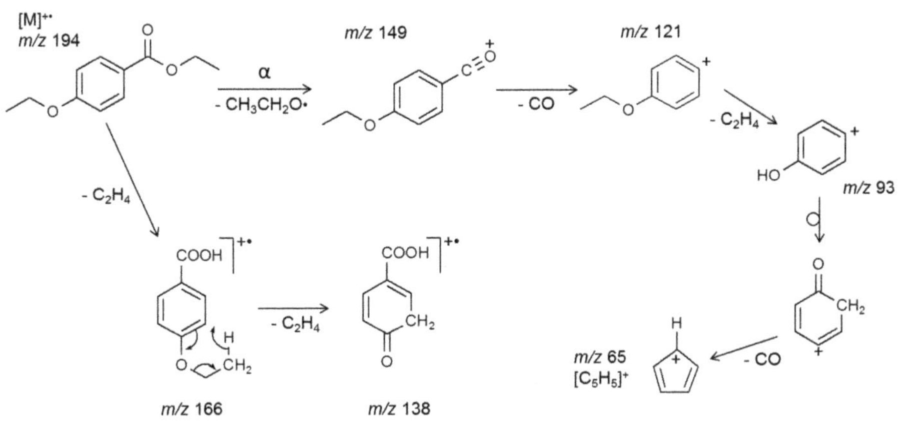

[M]⁺˙ → [M]$^{+\bullet}$

m/z 194   m/z 149   m/z 121

α
- CH₃CH₂O˙

- CO

- C₂H₄

- C₂H₄

m/z 93

HO

COOH   COOH

- C₂H₄

O   CH₂

m/z 166   m/z 138

m/z 65
[C₅H₅]⁺

- CO

O   CH₂

Scheme 7.30   EI-MS fragmentation reactions of the molecular ion of 4-ethoxy-benzoic acid ethylester, differing substantially from the pathways found for the isomer shown in Scheme 7.29.

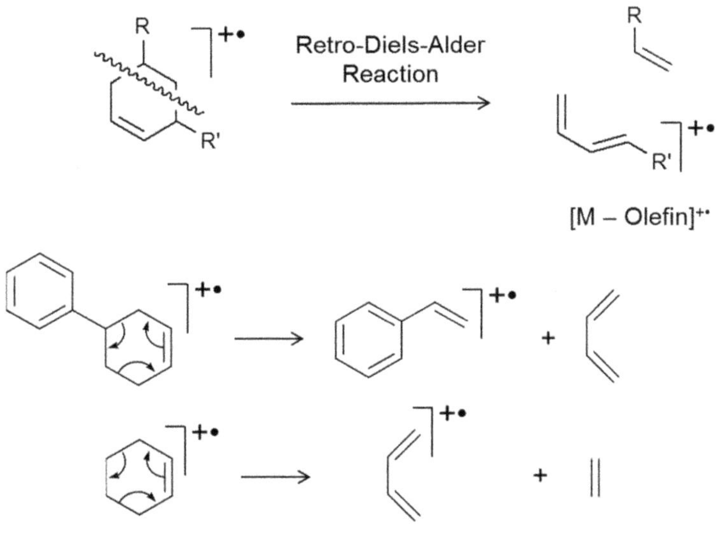

R

Retro-Diels-Alder
Reaction

R'

R

R'

[M – Olefin]⁺˙

+

+

Scheme 7.31   The RDA reaction divides a cyclohexene ring into a *diene* and an *ene-* moiety. The Stevenson and Audier rule is relevant for rearrangement reactions in which the OE⁺˙ character of the precursor ion is retained in the product ion species by loss of a neutral molecule, meaning that the radical cationic fragment with the lower IE will be formed.[6] In the case of cyclohexene this is butadiene, whereas in case of phenyl-cyclohexene it is styrene.

7.2.10   Retro Diels Alder Reaction

Organic molecules with a cyclohexene moiety can undergo the *Retro-Diels-Alder* (RDA) reaction to generate a diene and an alkene product species as illustrated in Scheme 7.31. Similar to the

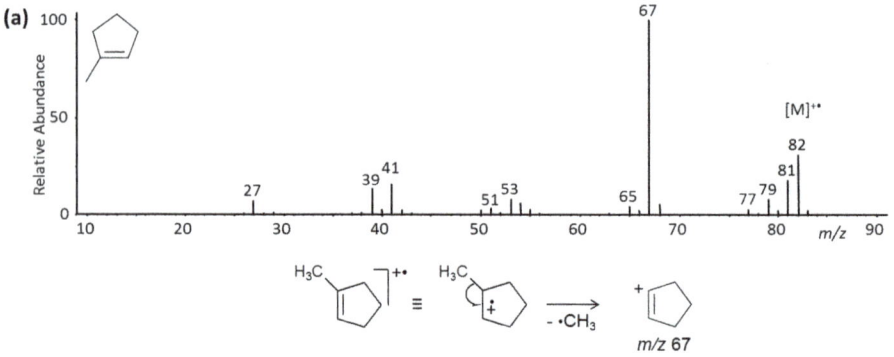

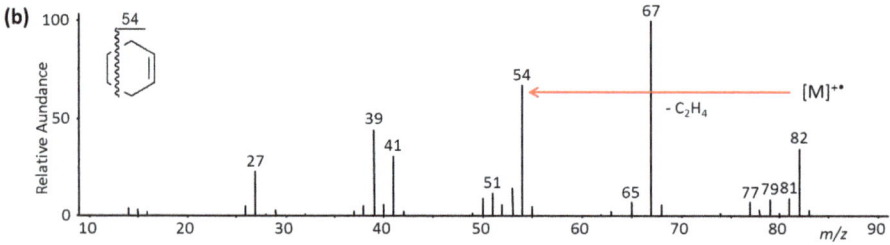

Figure 7.30 EI mass spectra of methyl-cyclopentene (a) and of cyclohexene (b). The predominant fragmentation reaction in the EI mass spectrum of methyl-cyclopentene is the loss of a methyl radical as shown below spectrum (a). The characteristic loss of ethane observed in spectrum (b) of cyclohexene *via* RDA (Scheme 7.32) explains the exclusive formation of the important fragment ion at *m/z* 54. This ion allows clear distinction between the two constitutional C_6H_{10} isomers depicted in (a) and (b). Spectra taken from the NIST Mass Spectral Library, Version 2.0f, 2008. Reproduced with permission.

McLafferty reaction the question is which of the two products of the RDA reaction will be observed in the EI mass spectrum and which will be lost as the neutral relies on the ionisation energy of the respective olefins (the Stevenson and Audier rule).[6] The RDA is important, especially for terpenes and other natural compounds with a cyclohexene ring system. In Figure 7.30 the EI mass spectra of methyl-cyclopentene and of cyclohexene are compared and the RDA fragment at *m/z* 54 and provides clear evidence for the RDA reaction and allows unambiguous structural assignment (see also Scheme 7.32).

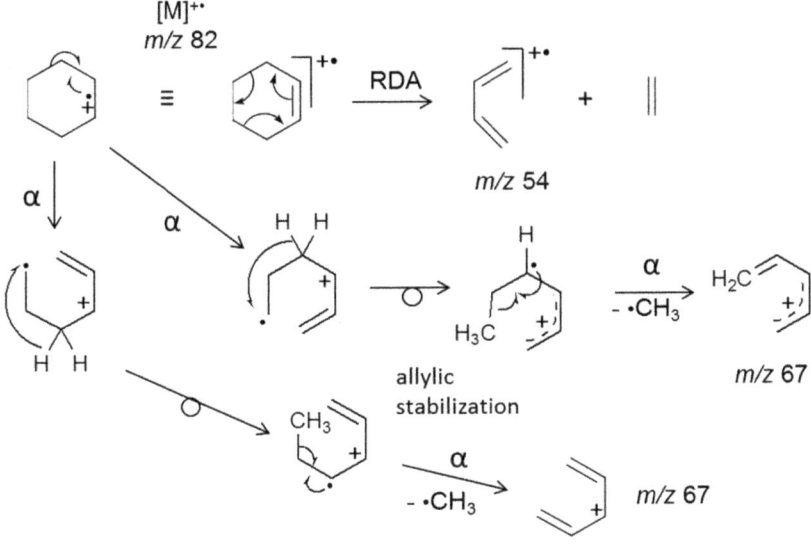

Scheme 7.32 Fragmentation reactions explaining the formation of the important ions in the EI mass spectrum of cyclohexene (see also the EI mass spectrum in Figure 7.30).

Further Reading

1. E. De Hoffmann and V. Stroobant, *Mass Spectrometry, Principles and Applications*, Wiley, Chichester, 2nd edn, 2002.
2. F. W. McLafferty and F. Turecek, *Interpretation of Mass Spectra*, University Science Books, Mill Valley, 1993.
3. H. Gross, *Mass Spectrometry – A Textbook*, Springer, Berlin, 2004.

References

1. H. Gross, *Mass Spectrometry – A Textbook*, Springer, Berlin, 2004.
2. W. McLafferty and F. Turecek, *Interpretation of Mass Spectra*, University Science Books, Mill Valley, 1993.
3. E. De Hoffmann and V. Stroobant, *Mass Spectrometry, Principles and Applications*, Wiley, Chichester, 2nd edn, 2002.
4. R. E. Ardrey, *Liquid Chromatography-Mass Spectrometry*, Wiley-VCH, Weinheim, 1993.
5. H. Budzikiewicz and M. Schäfer, *Massenspektrometrie eine Einführung*, Wiley-VCH, Weinheim, 6th edn, 2012.

6. W. Paul and H. Steinwedel, *Z. Naturforsch.*, 1953, **8a**, 448.
7. D. P. Stevenson, *Discuss. Faraday Soc.*, 1951, **10**, 35.
8. A. G. Harrison, C. D. Finney and J. A. Sherk, *Org. Mass Spectrom.*, 1971, 5, 1313.
9. E. Audier, *Organ. Mass Spectrom.*, 1969, **2**, 283.
10. S. Dewar, *J. Am. Chem. Soc.*, 1984, **106**, 209–219.

8 What Is Quantitative Analysis and How Do I Perform It?

Along with, "what is in my sample?" the next most important question is normally, "how much of something is in my sample?". In order to answer that question, the GC-MS must be setup and operated in such a manner so that:

- The system is sufficiently sensitive to detect the analyte and report amounts which meet the requirements for accuracy and precision.
- The quantitation is accurate, that is the amount reported is close to the true value.
- The quantitation is precise, so that the analysis is able to discriminate the variation seen in different samples to an acceptable degree.

The aim of this chapter is to address these needs. However, prior to addressing these points it is also important to understand how a response is generated by a GC-MS system and how that response is in turn measured. In common with many other chromatography-based systems, the signal from the detector is monitored by the analytical data system. There are several different manufacturers of GC-MS but they all measure the output from the GC-MS in a similar manner. Only single quadrupole MS (SQMS) is discussed here. In this case the

Gas Chromatography-Mass Spectrometry: How Do I Get the Best Results?
By Diane Turner, Mathias Schäfer, Steven Lancaster, Imran Janmohamed, Anthony Gachanja and Jason Creasey
© Diane C. Turner, Mathias Schäfer, Steven Lancaster, Imran Janmohamed, Anthony Gachanja and Jason Creasey 2020
Published by the Royal Society of Chemistry, www.rsc.org

collected signal is made up of the *mass-to-charge* ratio measurements and intensities as the mass spectrometer scans. See Chapter 5 for details of this mass analyser. The acquisition rate, mass range and mode of acquisition, either full scan or selected ion monitoring (SIM) for SQMS that is required will be determined by the analytical needs and is specified in the acquisition method (see Chapter 5). In full scan data acquisition, all ions across the specified range are acquired to generate a full mass spectrum for each corresponding chromato-graphic data point. This is useful for non-targeted analysis but results in a lower sensitivity. In SIM mode, in which 1–4 ions are acquired for each analyte, longer can be spent allowing each ion through to the detector resulting in a higher sensitivity.

Quantitation Ion: When quantifying, whether the data is from full scan or SIM acquisition, a single, unique ion (*i.e.* not present in any coeluting peaks) with a good signal-to-noise ratio should be selected as the quantitation ion. This will reduce the likelihood of interference from coeluting peaks, resulting in more accurate quantitation, espe-cially as integration should be baseline-to-baseline. The largest ion in the chromatogram is not always the best ion to select as, although it is abundant, it may not be as unique nor give such a good signal-to-noise ratio as possibly a higher mass but less abundant ion. It is always useful to try a few different ions and compare them with real samples containing coeluting matrix peaks.

Verification of Correct Identification: Along with the retention time, 1–3 qualifier ions are used to verify the identification from SIM data, by checking their ratio against the quantitation ion, usu-ally with a 20% uncertainty, although this can be up to 50% for small ions in complex matrices. From scan data, verification can also be achieved from the retention time and 1–3 qualifier ions as in SIM, but it can also be from comparison of the full mass spec-trum of the sample peak against the standard or library mass spec-trum for that analyte. Which of these can be used is dependent on the software.

In quantitative analysis the compound must be known. Quantita-tion is based upon comparison of the height or area of a peak with that of one or more standards of known concentration. Area and height vary linearly with concentration as long as the analysis system is chosen for the application, the methods are fully optimised and the conditions are carefully controlled and monitored with regular cali-brations, system suitability checks, the use of conditioning samples, and so forth. Samples and standards must be treated in exactly the same way for accurate quantitation.

Analyses on the Peak Height: The peak height can be measured with accurate precision. However, column temperature, flow rate and sample introduction techniques can all cause band broadening and therefore require careful control. Increased analyte concentration causing column overloading has little effect on the peak height leading to inaccurate results. As the column ages, band broadening and column bleed increase. Therefore, using peak height in a quantitative method is rare.

Analyses on Peak Area: The peak area is easy to determine with today's data analysis software. Peak area is largely independent of peak broadening effects, especially for reasonably large peaks. The peak must be well-defined for accurate measurements with 15–25 datapoints and a good signal-to-noise ratio (at least 10 : 1) improves accuracy even further with respect to the baseline. Therefore, peak area is the preferred data for quantitation as it is less sensitive to peak broadening effects.

8.1 How Can I Make Sure the System Is Ready for Quantitative Analysis?

Successful quantitative analysis requires the GC-MS system to operate in a way which generates a stable and repeatable output. Precise and accurate quantitation requires the GC-MS to be setup and maintained correctly and for the method to be properly optimised. The MS should be leak-free and tuned to ensure the highest performance and accuracy. The GC should also be well-maintained, clean and free of any activity. It is always recommended to check the GC-MS performance with a system suitability check. This is a term used to describe a test or tests which are made against pre-agreed acceptance criteria that (if properly designed) will illustrate whether the GC-MS is functioning to the required standard.

System suitability is designed on a method by method basis and is tailored to reflect the requirements of individual methods. However, there are a number of commonly used areas found in many system suitability checks. These are:

- chromatographic resolution;
- peak symmetry and tailing;
- signal to noise ratio.

Further details on the determination of these, along with other performance criteria, can be found in Chapter 6. Pharmacopeia such as the United States Pharmacopeia (USP) or European Pharmacopeia contain specific requirements for system suitability and these are a useful reference point.

After checking that the chromatography and sensitivity are good using a system suitability check, the repeatability of the method needs to be considered. When validating a method, this should be carried out as a minimum requirement.

8.2   How Do I Calibrate My System to Obtain Quantitative Data?

In order to determine how much of a substance is present in a sample, it is first necessary to ensure that you have a representative sample (see Chapter 1 for a detailed discussion on sampling), that is prepared and analysed using an optimised method.

The most accurate quantitation methods use a standard of the target analytes, prepared at a suitable concentration, to determine the response of each analyte in the detector at that concentration. This is because the detector response is not universal, that is, different substances respond differently within a GC-MS and give a different response per unit mass. The standards are prepared and analysed using the same method as the sample, the more closely the standard matches the sample analysis steps, the more accurate the result. The detector response of the analyte in the sample is then compared to that obtained for the analyte in the standard.

Semi-quantitation is used when a standard cannot be prepared for a target analyte, for example if the standard is not available, the analyte is an unknown or trends need to be tracked from sample to sample. The analyte is semi-quantified against the calibration curve of a calibrated analyte, analysed using the same method conditions. This is known as specific semi-quantitation and is usually used for both external and internal standard calibrations, as discussed later. The calibrated analyte should have similar properties, that is, belong to the same class and have a similar detector response if possible, the better the match the more accurate the estimation of concentration. However, the calibrated analyte probably does not have an identical response, an identical curve fit and range nor identical recovery or extraction efficiency in sampling or sample preparation.

Global semi-quantitation is good for analyses in which many compounds are present. It estimates the concentration against another compound, such as an internal standard, which may be quite different.

Specific quantitation will give a more accurate result than global semi-quantitation.

There are four major types of calibration that are typically used with GC-MS analysis:

- Normalisation
- External standard
- Internal standard
- Standard addition

Each method of standardisation has its benefits and disadvantages. Some of these methods use a response factor (RF), defined as the ratio between the concentration of the analyte and the response of the detector to that analyte. The most common method is to use the peak area:

$$RF = \frac{\text{Peak Area}}{\text{Concentration}} \qquad (8.1)$$

The most accurate and precise calibration is obtained when calibration occurs as part of the same batch of injections as the sample(s). It is however, for some applications using a certain type of calibration, possible to calibrate and then use that calibration for subsequent batches over the following days or even weeks, if the instrument is stable and does not change in response significantly. This runs the risk of not compensating for changes in response which might occur over time, therefore the system and method should be fully validated to ensure that this is not the case. Calibration should certainly be repeated if major changes occur, such as venting the GC-MS system or changing the GC column. For more difficult applications and certainly if complex samples are analysed, it is also good practice to inject a standard at intervals during a multi-sample run, to check that the response has not changed significantly. The interval must again be determined during method validation.

8.2.1   When Should I Use Normalisation as a Calibration Method?

In the normalisation method, the area of a sample analyte is calculated as a percentage of the total area of all peaks detected, excluding those from the solvent (or its impurities) above a pre-determined limit. Care must be taken when using this method, as it is semi-quantitative at best.

The normalisation approach makes the assumption that all compounds of interest have eluted in the chromatogram. You also assume that each peak in the chromatogram has a similar response. In GC-MS detection, the assumption around the response is unlikely to be true in the majority of instances, as at a minimum, different compounds will have different ionisation efficiencies resulting in different responses from the mass spectrometer. It would be particularly inappropriate to use this method if the mass spectrometer was operating in the SIM mode, as only a few ions are acquired per analyte rather than the full mass spectrum, therefore the response is very specific to the analyte. The total ion chromatogram (TIC) acquired in full scan mode is the most appropriate GC-MS data to acquire for this method.

An advantage of the normalisation approach is that no separate standards need to be prepared, therefore it is quick to obtain a result. The normalisation approach also compensates for variation in the injection volume and so forth, as all compared areas are in the same chromatogram. In order to be successful, all components must be within the linear portion of the detector response, eluted and detected. With the appropriate validation this approach might be appropriate, but careful consideration of the issues which may affect the response factors of each component detected is a factor that should be considered before this technique can be recommended for most GC-MS analyses.

8.2.2   When Should I Use External Standardisation as a Calibration Method?

The external standard method is the most straightforward method for quantitative chromatographic analyses. The concentration of the analyte is determined by comparing the response obtained in the sample solution to the response obtained with a standard solution.

The approach requires the preparation of one (single) or more (multi-level) standard solutions at concentrations around the approximate concentration of the target analyte. Chromatograms for the standards are obtained from the GC-MS. The analyte response (peak area) or relative response is plotted as a function of the concentration or relative concentration, ideally a straight line through the origin is obtained, although this is unlikely owing to background effects, and so forth. Chromatograms of the samples are obtained and the peak area of the analyte is compared against the calibration curve to determine the concentration of the analyte in the sample.

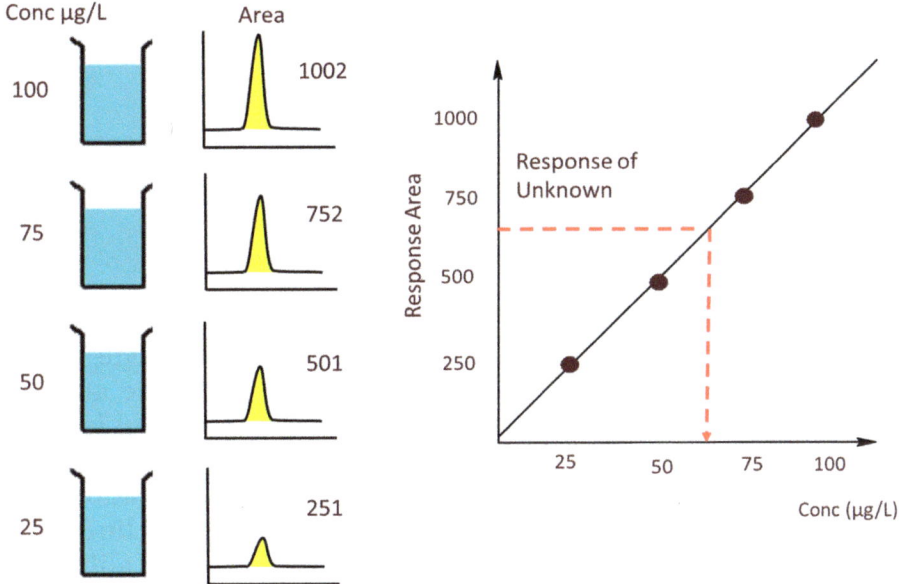

Figure 8.1 A schematic diagram of an external standard calibration process and curve. Courtesy of Anthias Consulting Ltd.

As an example, as illustrated in Figure 8.1, four standards of different concentrations such as 100, 75, 50 and 25 µg L^{-1} are analysed on a GC-MS system. A quantitation ion for the target analyte is chosen and integrated in each data file, then the peak areas are plotted in a calibration curve. Information should be entered into the software for the calibration curve to be generated.

- *Units:* it is best to use the units required in the report.
- *Curve fit:* linear, quadratic, cubic, logarithmic, exponential, and so forth, this depends on the detector's response to the analyte, try different types. In this example it is linear.
- *Origin (0,0):* if it should be ignored (recommended), included, forced or have a blank offset. Here it is ignored.
- *Weighting:* this is the importance of each point when fitting the curve. They can have equal weighting (shown here), or weight is 1/concentration (lower concentration has greater importance, 1/response), and so forth.
- *Correlation coefficient (R^2):* this is the fit of the points to the trendline, 1 = absolute fit. The requirement for fit depends on the analysis, for example >0.998.

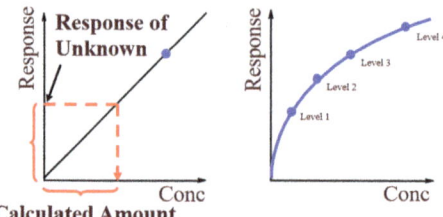

Figure 8.2 Single level (left) and multi-level (right) calibration. Courtesy of Anthias Consulting Ltd.

A calibration curve is produced for every target analyte to be quantified. If a single level calibration is to be used, the analyte must show a linear response to the detector and the origin is used as the second point, as shown in Figure 8.2 on the left. A multi-level calibration is used to cover a wider range of concentrations, as shown in Figure 8.2 on the right. It ensures the detector is linear (or non-linear) to the analyte across the concentration range calibrated and the analyte concentration should fall within this concentration range, otherwise it is not known if the curve is correct or not. A minimum of five calibration points is recommended to accurately determine the curve fit.

The external standard method (ESTD) is simple to set-up, but it tends to take time at first, then recalibration is automated. The sample typically has the same units as the standards, dilution and multiplier calculations can also be applied to amounts for reporting. Unlike normalisation, each peak is independent of the rest of the sample, therefore, only peaks of interest need to be separated, measured and calibrated.

In method development and calibration, you need to determine how many samples can be analysed before the system needs re-calibrating. Most analyses work on 20–40 injections per run including the calibration standards, check standards are analysed more frequently to ensure the instrument is still within calibration.

In the ESTD, the assumption is made that the samples are analysed in exactly the same way as the standards. This assumption is typically true but in GC it is very easy to vary, for example, the injection volume slightly with an air bubble, or if the detector response drifts, especially if the method is not fully optimised. This in turn means the amount reaching the detector will vary from injection to injection and results in errors. Therefore, an internal standardisation method is often used to compensate for these variations such as the injection volume. It is the most common kind of standardisation in use with GC-MS.

8.2.3 When Should I Use Internal Standardisation as a Calibration Method?

The highest precision is obtained using the internal standard method (ISTD), as low as 0.5–1%, however this is application dependent. The ISTD compensates for slight differences in injection volume, flow rate, other GC parameters and detector drift. It normalises all analytes against the peak from an internal standard.

An accurate amount of an internal standard (IS), at a fixed concentration, is added to every sample, calibration standard, and so forth in equal amounts, usually just before analysis. Additional standards can also be added before any sample preparation steps, to check for problems, in environmental analyses these are called surrogates. For example, these are used to check for:

- Losses during volumetric or gravimetric manipulation.
- Losses owing to evaporation or to surfaces during storage or preparation.
- Various other potential losses during sample preparation, such as the extraction efficiency and matrix interference in the preparation of that particular sample.

The point at which the IS is added is dependent on what it is being used to check for – just the GC-MS or sample preparation too, and if so which steps? It is worth using a different compound in each step, to monitor and identify the source of potential problems, but only one can be used to normalise each target analyte peak! Usually, surrogates are added before sample preparation and are quantified in the same way as the target analyte, they are also normalised to the internal standard. Internal standards are usually added to the chromatography vial before placing into the autosampler.

A calibration curve is created, very similar to that in the ESTD. However, it is now the response ratio (or relative response) between the internal standard and the target analyte in the calibration standard against the concentration ratio (or relative concentration) that are plotted. Then, to determine the concentration in the sample, it is the response ratio from the sample chromatogram that is used, therefore the IS must always be found.

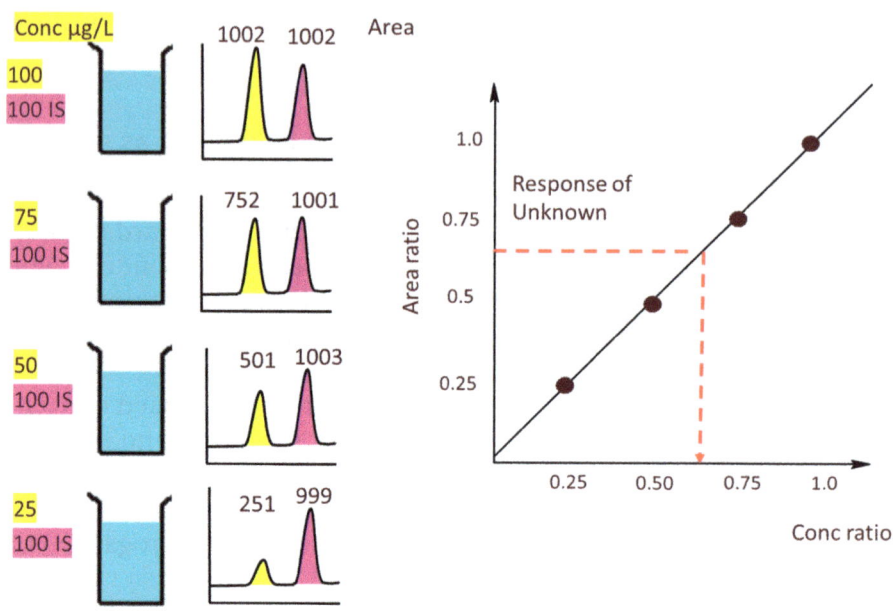

Figure 8.3 A schematic diagram of the internal standard calibration process and curve. Courtesy of Anthias Consulting Ltd.

As illustrated in Figure 8.3, four standards of different concentrations such as 100, 75, 50 and 25 $\mu g\ L^{-1}$ are analysed on a GC-MS system. A TIC or single ion is chosen, the target peak and internal standard areas integrated and their relative ratios determined, the peak area ratio is then plotted against the concentration ratio. When an unknown sample is then analysed, the response ratio of these two peaks is then used in the calibration curve to determine the concentration ratio and, as the internal standard concentration is known, the target analyte can then be determined.

A simplified procedure for quantitation using internal standards requires the calculation of the relative response factor (RRF), see eqn (8.2), for the target analyte peak in the calibration standard from the RF calculated using eqn (8.1), for the target analyte (A) against the internal standard (IS):

$$RRF_A = \frac{RF_A}{RF_{IS}} = \frac{Peak\ area_A}{Concentration_A} \bigg/ \frac{Peak\ area_{IS}}{Concentration_{IS}} \tag{8.2}$$

Then, in the sample the concentration of the target analyte (A) can be calculated using:

$$Concentration_A = Peak\ area_A \times \frac{Concentration_{IS}}{Peak\ area_{IS}} \times \frac{1}{RRF_A} \qquad (8.3)$$

The ISTD corrects for variations in the injection volume, detector response, and so forth and also corrects for long-term response changes (days). However, the use of an IS adds more peaks to the chromatogram to resolve, detect and to set-up in the method, adds more time for sample preparation (consider automation in respect to both time and accuracy) and can be difficult to find the most appropriate IS (especially in large mixture of analytes), but the pros usually outweigh the cons.

8.2.3.1 How Do I Choose a Suitable Internal Standard?

Internal standards are selected to match as closely as possible the physical and chemical characteristics of the target analyte(s), including the detector response factors, potential activity in the system, volatility, and so forth, so that any slight variations in the run will have the same effect on both the internal standard (IS) and the analyte(s). IS compounds must not already be present in the sample, should not react in the sample or degrade in the GC (unless the target analyte does as well), should be distinguishable from other analytes with either chromatographic or spectral resolution and should be easy to integrate accurately.

An IS should typically elute very closely before or after the target analyte(s) for which it is being used as an IS. A simple chromatogram, with few analytes, will usually have one IS eluting near the middle of the target analytes. More complicated chromatograms with a greater numbers of target analytes and/or different classes of analytes should use multiple IS compounds: one or more for each class of compound, or multiple internal standards to cater for volatility, eluting at the front, middle and end of the chromatogram.

Internal standards should be added to the calibration standards and samples at the same concentration and at concentrations which are approximately equal to that expected for the target analyte (certainly within an order of magnitude). For multi-level calibrations, they should be at approximately mid-concentration, however if the target analytes are more likely to be seen at lower concentrations, particularly if they are active compounds, then their concentration should reflect that.

The average RRF ratio of an IS to a target analyte is expected to have a smaller standard deviation than a RF calculated only for the external

Table 8.1 Set of calibration data.

Name	Response of internal standard	Response of analyte	Amount analyte	Response ratio
Cal 1A	12409	3117	10.00	0.2512
Cal 2A	12886	6725	20.00	0.5219
Cal 3A	12879	13686	40.00	1.0627
Cal 4A	11817	15819	50.00	1.3387
Cal 5A	12728	31902	100.0	2.5064
Cal 1B	12812	3592	10.00	0.2804
Cal 2B	13296	7419	20.00	0.5580
Cal 3B	12793	15102	40.00	1.1805
Cal 4B	13306	18567	50.00	1.3954
Cal 5B	13326	37382	100.0	2.8052

standards. Table 8.1 shows the data for an analyte standard prepared at five different concentrations, each injected twice into a GC-MS.

The response of the internal standards is plotted in Figure 8.4(a) together with the average response. It can be seen that Cal 1A and Cal 4A are less than average and Cal 2B, Cal 4B and Cal 5B are higher than average. This variation affects both the IS and target analyte. Without the IS normalising the data, this affects the quality of the data (a correlation coefficient of $R^2 = 0.9827$ without IS as shown in Figure 8.4(c), $R^2 = 0.9914$ with IS as shown in Figure 8.4(b)).

8.2.3.2   I Don't Know What Is in My Sample So What Can I Use as an IS?

In environmental or biological samples for example, any chemical could potentially contaminate them and therefore it is very difficult to choose an IS which will definitely not be in the sample. Therefore, isotopically labelled internal standards are often used, these have both chemical and physical properties that are similar to the target analyte. They may or may not be chromatographically resolved, indeed, the lack of chromatographic separation is an indication that the physiochemical properties are very closely matched, but they can always be analytically separated from the non-deuterated analyte owing to differences in their mass.

Isotopically labelled IS are the deuterated analogues of the target analytes, some of the hydrogens in the molecule have been replaced by deuterium. Deuterium, which has an extra neutron compared to hydrogen, increases the molecular weight of the analyte in question by one Dalton for each deuterium atom added to the molecule, allowing a mass spectrometer to resolve the two owing to their difference

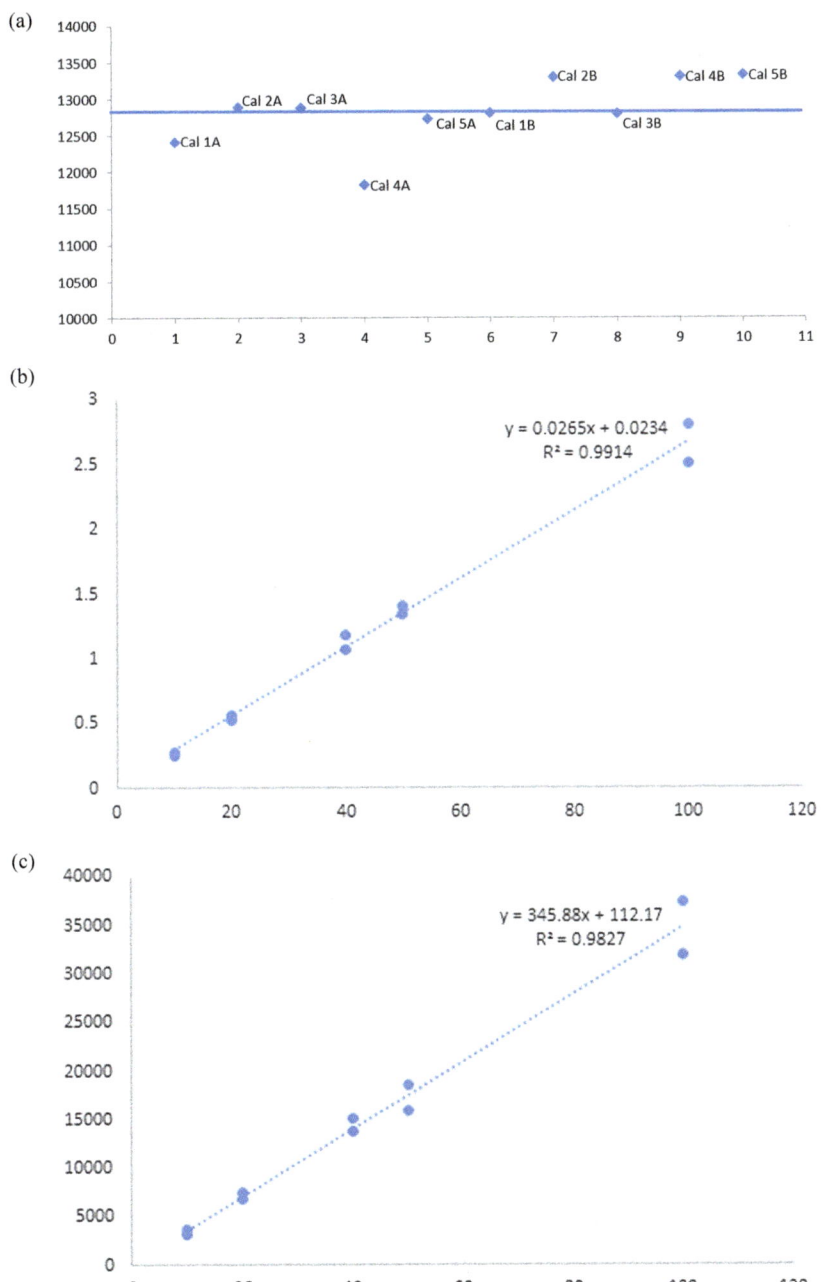

Figure 8.4 The response of the internal standards, including the average response (a), the analyte response ratio (b) and analyte response (c).

in mass. This deuteration is achieved during the synthesis of the molecule by the use of deuterated analogues of pre-cursors of the target analyte. This means that the availability of a deuterated IS is limited by the ease of production during synthesis and thus can be expensive to produce. However, the long-term costs in terms of improved accuracy should be strongly considered.

ISTD quantitation using deuterated internal standards is performed in exactly the same way as for non-isotopically labelled IS, however the retention time of the IS and target analyte will probably be the same and care must be taken to choose a quantitation ion that is unique to each, rather than a low-mass ion that is common to them both if fragmentation has removed the deuterated part of the molecule.

A special type of calibration that utilises isotopes as internal standards is isotope dilution analysis. This gives very high levels of accuracy and precision, but this is beyond the scope of this book. The reader is referred to Guidelines for Achieving High Accuracy in Isotope Dilution Mass Spectrometry (IDMS), in the further reading section.

The example presented in Table 8.2 uses deuterated polycyclic aromatic hydrocarbons (PAHs) to monitor trace quantities of PAHs in a complex matrix. Table 8.2 shows how multiple deuterated PAHs provide a method to match each PAH with its deuterated analogue to achieve precise quantitation.

8.2.4   When Should I Use Standard Addition as a Calibration Method?

Standard addition is a quantitative analysis approach in which the calibration standard is added directly to aliquots of the sample, rather than being analysed separately, as in ESTD. Multiple (a minimum of three) aliquots of the sample are taken and to each one a different amount of a known concentration of the calibration standard is added as a spike, to one aliquot no standard is added. The aliquots are then analysed as a sequence of injections and the response of

Table 8.2   Target PAHs and their corresponding deuterated internal standards.

Polycyclic aromatic hydrocarbon	Monitored SIM ion	Internal standard	Monitored SIM ion
Naphthalene	128	Naphthalene-d_8	136
Pyrene	202	Pyrene-d_{10}	212
Benzo(*e*)pyrene	252	Benzo(*e*)pyrene-d_{12}	264
Benzo(*ghi*)perylene	276	Benzo(*ghi*)perylene-d_{12}	288

the target analyte(s) is measured in each. The calibration curve is then constructed, similar to ESTD, in which all responses are plotted against the concentration added. An extrapolation can then be made to determine the amount present in the un-spiked sample.

This process is typically done *via* the fitting of a regression line to the solution responses and extrapolation back through the non-zero *Y*-intercept to establish the concentration present when no quantity of additional standard has been added. From this you can calculate the amount present in the sample.

Standard addition, as a method of calibration, is slightly more complex than the methods described earlier and as such takes a little more time and care to accomplish. It does however offer advantages over other methods, particularly in which the sample matrix has a noticeable effect on the accuracy of the results, known as the matrix effect. The standard addition method adds a known amount of the target analyte to the sample matrix and thus the influence of the sample matrix is built into the analysis. It is also possible to account for sample matrix effects in the other calibration methods, but this method does this most directly and accurately.

The standard addition method clearly requires a relatively large amount of sample compared to other methods as it is required to make multiple measurements, the sample must also be homogenous. A new calibration curve must also be constructed for each sample to obtain accurate results. Some GC-MS software may not have this method of calibration built-in and it may be necessary to calculate the amount present from a regression equation, produced from statistical software or a spreadsheet program.

8.3   How Do I Validate My Method?

Validation is a key step to ensuring a method is able to perform qualitative or quantitative analysis. Validation may be a quick couple of experiments to check that a method will give the correct results and should be performed for every method, even those used in research rather than routine analysis. Validation tends to be more in-depth for quantitative methods and the extent depends on the industry. For example, methods for the routine, environmental analysis of water undergo validation using five different types of water matrices, with at least a 6-point calibration curve, analysed in a minimum of 11 separate batches of duplicates with around 69 injections per batch across a minimum of 6 days.

Without a formal validation the user of a method is unable to demonstrate that the results from the method are valid. The validation seeks to ensure the method is fit for its intended purpose and therefore it is important when thinking about how the method will be validated to consider what the requirements are for the method. Given below are some general standards and points to consider when validating a method. However, it is important to agree a specific validation protocol before starting the experiments which make up the validation. This should be discussed with the users of the data you are about to generate, and you should seek to define exactly what problem statement the analysis seeks to address. This protocol should agree the elements of the validation and the acceptance criteria that you will apply to the results. A failure against the pre-agreed acceptance criteria will mean your method is not valid and will need more development to understand and fix the issues seen before the validation is repeated and the method used for 'real' sample analysis. Therefore, validation should not be untaken until you are confident you have developed a method that can meet the acceptance criteria set. Some of the elements described below can be used during the development of a method to test if the method is ready for validation, but it is good practice to have a separate documented formal validation process once the method is ready.

Some performance characteristics in method validation are:

- Linearity
- Limit of detection
- Limit of quantitation
- Accuracy
- Precision
- Robustness

8.3.1   Linearity and Calibration Curves

When using standards to determine the amount of an analyte present it is very common to construct calibration curves to measure in a sample the response of these standards over a range of concentrations. Linearity is the ability to obtain results which are directly proportional to the concentration or amount of analyte in the sample within the given concentration range (minimum 3–5 points). It is generally expected that the variation in analyte concentration will have a linear relationship with respect to the response given by the MS detector and hence a linear curve fit is the most common. There are two reasons why the response may not be linear:

- Operation of the detector is outside its linear dynamic response range.
- Adsorption of the analyte in the autosampler or GC-MS system, particular polar analytes at lower concentrations. This is discussed further in later chapters.

8.3.1.1   How Does Dynamic Range Affect Linearity?

Operating outside the linear dynamic range of the detector normally occurs if the sample is at too high a concentration (Figure 8.5). The exact upper end of the limits must be determined by experimentation, but it becomes clear if the chromatographic peaks are flat-topped that the detector is saturated. The relative intensities of the fragment ions in the mass spectrum are also likely to be affected which may mean that library searching is not successful or gives the wrong result. However, for target analysis, the TIC may appear saturated, but extraction of the quantitation ion, if unique, may not be. All ions used, including the qualifier ions, should be checked. The working linear dynamic range is expected to be between 3–4 orders of magnitude, but this is very dependent on the analytes and the GC-MS model itself. In isolation, the MS linear range is much wider than this but the effect of adsorption throughout the system has a significant bearing on the overall linear range.

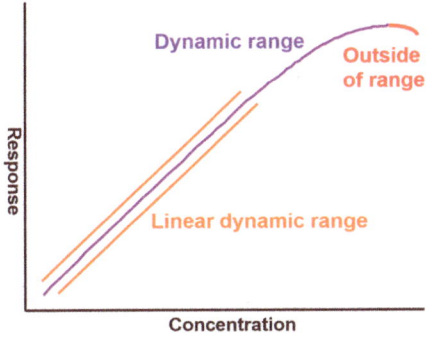

Figure 8.5   Dynamic range. Courtesy of Anthias Consulting Ltd.

Although less common, it is sometimes necessary to fit a non-linear curve to the detector output to achieve acceptable quantitation. This may be a way to fit the data observed to the calibration curve. It should only be done if there is no way to resolve the problem to achieve a linear fit, but it could be necessary when operating across a wide range or operating close to the limit of detection in complex matrices.

8.3.1.2 What Is Regression Analysis, and When Should I Use It?

A commonly used statistical technique is regression analysis. The process of linear regression can be applied to the calibration data to determine a "line of best fit". The linear regression line is then used to predict the values of unknown samples. Linear regression can be performed either within the software of the GC-MS or using a statistical function such as that built into Excel™.

The output from linear regression is an equation for a straight line which represents a calculated slope and intercept which is the "best fit" to the data. Along with the slope and intercept, a statistical value known as R^2 (correlation coefficient) is calculated which compares the estimated and actual values. The closer these values are, the closer this value is to 1. This coefficient is a useful check on the quality of the calibration data. It is common practice to set a minimum value of R^2 for the method and use this as a system suitability check. Similar information is provided when using other curve fits, such as quadratic.

8.3.2 How Do I Calculate the Limit of Detection?

The detection limit (DL) or limit of detection (LOD) has multiple definitions, including:

- It is an estimate of the minimum amount of a substance that can be distinguished from a blank value that an analytical process can reliably detect.
- It is an analytical signal from a standard that can reliably be distinguished from "analytical noise", with a signal to noise ratio (SNR) of 3:1.

The LOD is analyte- and matrix-specific and is laboratory dependent. LOD is estimated from the mean of the lowest concentration of a known standard and the standard deviation of the mean, and is reported with a confidence factor. Below this limit, the system is considered not able to provide adequate detection of an eluted substance. It is normally calculated at the same time as the limit of quantitation (LOQ), (see the next section).

Related to the LOD is the:

- Instrument detection limit (IDL) which is the lowest concentration that an analytical instrument can measure with a SNR = 3.
- Method detection limit (MDL) which is the minimum concentration of an analyte that can be measured and reported with 99% confidence that the analyte concentration is greater than zero. It is determined from the analysis of a sample in a given matrix type containing the analyte which an analytical instrument can measure after the necessary sample preparation, for example solid phase extraction (SPE).

Usually, the LOD is the reported MDL, as it should be determined from the entire sample analysis method. It is a property of the analytical procedure, sample matrix and measurement system. It is a statistic, an estimate that includes both systematic and random method errors.

To calculate the MDL, first determine the LOD for each target analyte from the lowest concentration standard at which the peak can still be detected with an SNR of at least 5:1 (see Chapter 6 for details on SNR). Prepare and analyse through the entire analytical method procedure seven or eight replicates at this concentration. Calculate the concentration for each replicate of each analyte against the calibration curve. Calculate the mean concentration and standard deviation for each analyte. Finally, multiply the standard deviation by the Student's t-value at 99% confidence and $n-1$ degrees of freedom, for 7 replicates = 3.143, and for 8 replicates = 2.998.

8.3.3 How Do I Calculate the Limit of Quantitation?

The LOQ, practical quantitation limit (PQL) or reliable quantitation limit (RQL) is the smallest measured content of an analyte above the LOD which can be made with a specified degree of accuracy and precision (*e.g.* relative standard deviation (RSD) 20% and accuracy of 75–125%). It is considered to be an SNR of 10:1 or can be determined as, for example, 3 × LOD or MDL values.

From the LOQ, the limit of reporting (LOR) or reporting levels can be determined. This is usually defined by a regulatory agency, company or group for more accurate quantitation results and is usually higher than the LOQ, for example 2 × LOQ. The relationship between the LOD, LOQ and LOR can be seen in Figure 8.6.

An example calculation of LOD, LOQ and LOR is shown in Table 8.3 for benzene in drinking water, with a suggested LOD of 3 µg L^{-1}. A calibration curve was created and seven replicates analysed ($n = 7$) of a 5 µg L^{-1} standard which were all analysed under the same conditions.

8.3.4   What Is the Difference Between Precision and Accuracy?

The precision of a measurement can be defined as the variation seen between replicate determinations of the amount of an analyte in a sample, and the closeness of agreement in the same homogenous

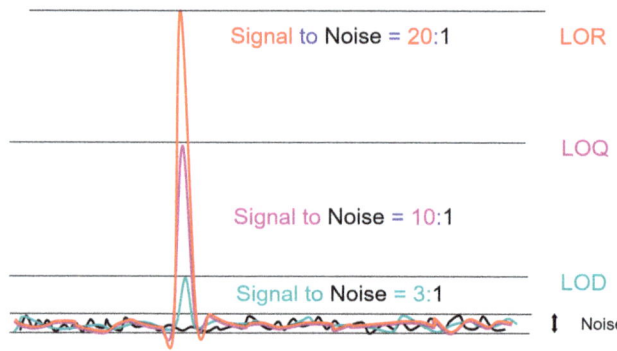

Figure 8.6   'Relationship between LOD, LOQ and LOR. Courtesy of Anthias Consulting Ltd.

Table 8.3   Example calculation of LOD, LOQ and LOR.

	Value
Concentrations measured (µg L^{-1})	7, 5, 5, 6, 5, 4, 5
Mean conc.	5.29
Std deviation	0.951
Statistical confidence (99% students *t*-test)	$n-1$: 6 = 3.143
	$n-1$: 7 = 2.998
	$n-1$: 8 = 2.896
LOD	0.951 × 3.14 = **2.9 µg L^{-1}**
LOQ (*e.g.* 3.3 × LOD)	2.9 × 3.3 = **9.9 µg L^{-1}**
LOR (*e.g.* 2 × LOQ)	9.9 × 2 = **19.7 µg L^{-1}**

sample. Good precision is desirable as this means the amount reported for a given sample will be consistent. Accuracy is defined as the closeness of the result to its true value, accuracy is obviously desirable as you want to produce a result which is representative of the true value. Ideally a result would be both accurate and precise. Both accuracy and precision are determined during method validation.

8.3.5   How Do I Determine Accuracy?

Accuracy is a key property needed for quantitative analysis. It is a measure of the "trueness" of the method. That is how close the reported result is to the actual amount present in the sample. As this is one of the prime reasons for conducting the analysis it is very important to find a suitable way of determining how accurate the method is during validation.

8.3.5.1   What Methods Can I Use for Measurement of Accuracy in Samples?

There are multiple methods for determining accuracy, however the most common is referred to as the "spiking and recovery" method. This method determines accuracy by the addition of a known amount of the target analyte to the sample (spiking), the sample is then analysed with at least seven replicates and the concentration of each is calculated from the calibration curve. This result is then compared to the spiked amount and the recovery or accuracy reported as a percentage. The example outlined in Table 8.4 is for a standard at 150 µg mL^{-1}, with the analysis of seven replicates.

Table 8.4   Example of an accuracy calculation.

Sample	Peak area	Conc.[a]	%Recovery[b]
1	173 865	148.8	99.2[c]
2	174 926	149.7	99.8
3	172 933	148.0	98.6
4	173 456	148.4	98.9
5	179 557	153.6	102.4[c]
6	176 425	150.9	100.6
7	174 536	149.3	99.6
Mean		**149.8**	

[a]From the calibration curve';
[b]%recovery = conc./known conc. X 100%;
[c]accuracy of 99.2–102.4%.

The expectation is that the result will agree with the amount added, that is 100% recovery. Whilst this is a simple concept in practice the spiking and recovery experiment has many pitfalls and complications and needs to be conducted carefully to avoid an unexpected result. It is recommended to always spike at three different levels of concentration, for example low, mid and high, as the recovery may vary depending on the concentration.

8.3.5.2 *Do I Have a Sample Which I Can Successfully Spike?*

A blank sample for spiking is required that is representative of the sample type, that is, it is the matrix but is free of the target analyte(s). You may be in a position to manufacture a sample which has the target either removed or not present. If that is not possible, then by analysing the sample with and without a spike it is possible to account for the amount present. Conducting a spiking and recovery experiment is much harder if the target analyte is present and needs to be subtracted from the result.

Additionally, adding a very small, known amount of the target analyte(s) to the blank sample accurately can be problematic. Typically, it may mean adding the analytes in a solution in order to achieve the required concentration in the blank sample. It also may mean adding volatile or reactive analytes to a matrix in which it is easy to lose all or part of the spike and the sample matrix will not allow complete recovery. Issues around solubility and heterogeneity within the blank sample also need to be considered. Usually, a solvent is used that is GC-amenable, mixes well or dissolves in the sample and one that does not react with the sample, for example methanol spikes.

Hopefully the method has been optimised to ensure the target is completely recovered, but if this is not robust or reliable this may be reflected in poor recoveries of the spiked analyte.

When quoting results it is always good practice to calculate the confidence intervals of the mean. This is the range within which the true value lies (assuming that only random errors are present and any systematic errors are totally absent).

8.3.6 How Do I Measure Precision?

Precision is a measure of how reproducible the result is for the target analyte. Good precision is considered to be one of the key parameters in quantitative analysis. Without it the method is unlikely to give a reliable, consistent result which is a key requirement in quantitative analysis. Owing to this key requirement, precision is often broken down into several sub-categories during validation to establish whether each element is functioning correctly and to determine which of these is the largest source of error. Precision can be considered on three levels:

- Repeatability, such as short time intervals (intra-day), it is useful to sub-categorise into:
  o Instrumental precision
  o Standard precision
  o Sample precision
- Intermediate precision, such as different days, analysts or systems.
- Reproducibility, for example different laboratories or inter-laboratory studies.

Generally speaking, it is expected that the variation will increase as we move down the list as the number of sources of potential error increases. This is therefore reflected in the acceptance criterion that should be set for each of these. The highest being for repeatability and the lowest for reproducibility.

Repeatability looks at the variation in measurements on replicate samples analysed using the same single instrument, with the same method and the same analyst performing the measurement. Reproducibility measures the entire study, with different analysts, different instruments and at different laboratories through controlled interlaboratory test programs.

Precision is expressed as the standard deviation (σ), variance (σ^2) or coefficient of variation (%RSD). The normal distribution of results can be seen in Figure 8.7, in which the mean is in the centre (μ) and 1σ = 68.26%, 2σ = 95.44% and 3σ = 99.730%.

8.3.6.1 How Do I Measure Instrumental Precision?

Instrumental precision is defined as the variation owing only to the system (autosampler (injection only, no sample preparation), inlet, column and detector). It is looking at the variation in peak response and is a repeatability experiment, as the replicates are analysed intra-day. This measurement is typically made by calculating the relative standard deviation of the detector response, usually the peak area or occasionally the peak height. The format of this measurement is

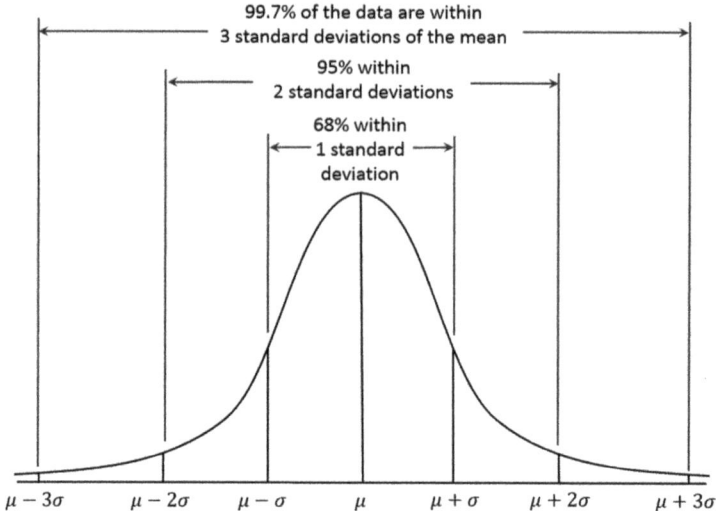

Figure 8.7   Normal' distribution. Reproduced from ref. 1 [https://commons.wiki-media.org/wiki/File:Empirical_Rule.PNG]underthetermsofaCCBY-Sa4.0license [https://creativecommons.org/licenses/by-sa/4.0/deed.en.

typically made through replicate injections of a mid-point calibration standard solution with enough replication to be statistically valid, typically a minimum of seven. From these replicate injections, a measurement of the standard response (area or height) is made and the standard deviation and mean are calculated. From these the percentage RSD (%RSD), also known as the coefficient of variation or the variation coefficient, is calculated as:

$$\%\text{RSD} = \frac{\text{Standard Deviation}}{\text{Mean}} \times 100\% \tag{8.4}$$

An example calculation is shown in Table 8.5 for the determination of a mid-point calibration standard at 50 μg mL^{-1} with seven replicate injections.

The expectation is that the variation in the responses seen here is low. How low will depend on your method and the concentration range you are working at. When expressed as the %RSD, at high concentrations the variation could be as low as 0.1% and at trace concentrations perhaps at high as 5–10%. Typically, good precision for GC-MS analysis is 3% RSD, although this may be improved by the use of internal standards, which should be used in the calculations if used. It is useful to do these calculations both with and without the use of internal standards, to check how good the method really is and how much the

Table 8.5 Example of repeatability.

Injection	Peak area
1	173 865
2	174 926
3	172 933
4	175 011
5	179 557
6	176 425
7	174 536
Mean	**175 099**
SD	**2204**
RSD	**1.26%**

IS is needed to improve the precision. If you cannot obtain good precision with a calibration standard, then method development should be revisited, as it will only get worse with real samples!

From these results you should be able to conclude that the system setup (injection, chromatographic separation and detection) is functioning correctly and able to produce a consistent result.

8.3.6.2   How Do I Measure Standard Precision?

Standard precision is defined as the variation owing to the preparation of standards and analysis of the standards used in the method. This is calculated in an experiment in which multiple weighings of a standard are made then injected repeatedly into the GC-MS.

Typically, three separate weighings are made and each of these standards is injected at least three times making a total of nine injections. The responses obtained from the GC-MS analysis of the standard analytes analysed under the set method conditions are used to assess the method.

The overall average response for each analyte in the method, together with the standard deviation observed can be used to calculate a %RSD, as shown in Section 8.3.6.1. A variation of around 1–5% is considered acceptable, once again trace analysis is expected to give a higher variation. Other experimental protocols are acceptable but it is important to focus on variation owing to the preparation and analysis of standard solutions.

From these results you should be able to conclude that the standard preparation (weighing, volumetric dilutions, and so forth) and their analysis is acceptable.

8.3.6.3   How Do I Measure Sample Precision?

Sample precision is defined as the variation owing to the preparation of samples and analysis of the samples. This is calculated in an experiment in which multiple replicate weighings/preparations of a sample are made then injected repeatedly into the GC-MS.

This is the first experiment in which the output will be a calculated result rather than simple measured responses from the GC-MS analysis. Therefore, it is important to ensure that the standard variation is acceptable before you conduct this experiment as it will have an influence on the result.

This experiment is principally designed to monitor precision of the sample preparation and detection and not the variation from sample batch to sample batch. Therefore, it is better to use a single batch of sample to perform this experiment to isolate the variation observed. As with standard precision, this experiment will study variation owing to weighing and any other manipulation performed on the sample such as volumetric dilution. It will also study any other sample specific parts involved in the sample preparation, for example sample extraction, sample clean-up or sample derivatisation.

The principle is the same as for standard precision, multiple replicate preparations (a minimum of seven is recommended) should be made (as per the method) each of these preparations are then run in a sequence of injections including the standards. The amount in the samples is then calculated as per the method and the average amount and standard deviation calculated. From this measurement a %RSD can be calculated and used to judge the acceptability of the calculated results.

Once again acceptable %RSD will vary owing to the exact requirements, but precisions between 1–10% are considered acceptable. A precision of 10% is typical for trace analysis and is dependent on the application, with higher values expected for some analytes, particularly those in complex matrices such as food samples.

8.3.6.4   How Do I Measure Intermediate Precision?

One of the final steps, depending on if the sample analysis using the method will take place in one lab or multiple labs, is aimed at measuring variables which might affect the precision when the method is being used routinely. The aim of this experiment is to introduce variations that mimic routine operation, therefore the exact design will vary a lot, depending on the use of the method in the laboratory, for example will different instruments or analysts be involved in this application? As a minimum it should involve a second analyst in the preparation of standards and samples and possibly the analysis.

It could include:

- different analysts;
- different analytical instruments;
- the same instrument but with a different column and consumables;
- different day of analysis

Regardless of the exact experiment protocol, the aim of this experiment is to determine that the variables introduced have not had a detrimental effect on the precision. To measure this a repeat of the sample precision is made including the variations listed above. A statistical analysis can then be made to compare the results. This can be as simple as determining the overall %RSD of all the measurements or a more in-depth analysis such as ANOVA (analysis of variance) or other statistical methods. For further information the reader is referred to the reading list at the end of this chapter.

If a simple overall %RSD is used, you would expect this variation to be between 10–25%, 25% being appropriate for a trace analysis method.

8.3.6.5 Reproducibility and Inter-laboratory Studies

To fully assessment the robustness of a developed GC-MS method, there is no greater test nor way of improving method confidence than to use that method in other laboratories and then assess how the results generated within those laboratories compare with yours. It determines the ability to replicate the same results as others. This is known as an inter-laboratory or round robin study.

It is self-evident that this introduces the widest range of variables into the process and therefore is the ultimate test in determining the quality of the methodology. If conducted in a systemic way, it will give the most accurate determination of the true value of an analyte(s) in a sample. It requires careful experimental design if meaningful results are to be generated, which includes a careful statistical analysis of the output. One might expect variation in the result to come from many sources and thus the statistical analysis conducted should be sufficiently rigorous to identify where variation occurs and account for that in the final declared result.

This type of analysis relies on the tight control of as many variables as possible, whilst recognising that the nature of the test will introduce variation. Therefore, the experimental design may explore a number of factors. The full method could be exactly the same in all laboratories, from sample preparation through to data analysis, or the analysis method itself might be different. Some of the variables introduced in inter-laboratory studies include:

- different methods of sample preparation;
- different instrument configurations and types;
- different environmental conditions;
- different interpretation of the method instructions;
- different standard sources.

Therefore, part of the experimental design will be to accept the variation or attempt to control it. A detailed pre-agreed experimental protocol is very important in establishing the agreed ways of working and standards to adopt. For example, the protocol may specify reporting units and style and the number of replicates to introduce into the analysis. Statistical analysis *via* ANOVA is a collection of statistical models used to analyse the differences among group means and their associated procedures, that is the "variation" among and between groups.

Thus, with this type of analysis we are able to see if the variation we see in sample determinations is less or more than we expect from the analysis. It is clearly important to understand if differences we see between samples are truly a reflection of differences between samples or an artefact of the analysis.

Depending on the requirements of the analysis, more or less time should be spent on establishing the variation owing to the analysis method, and the repeatability. As a minimum, replicate analysis of samples provides a method of verification that the result obtained for a given sample is valid.

8.3.7 How Do I Validate Standard Stability?

During the validation of a method, it is necessary to demonstrate that standards employed by the method are fit for purpose and remain so for the duration of the analysis. If standards are prepared in advance of the analysis, which is often the case, it is important that standards remain unchanged with respect to the concentration and response throughout their use as standards. Standards may be prepared at a

Table 8.6 Example protocol for standard validation.

Standard description	Storage conditions temperature & light	Storage time	Example protocol
Stock solution	2–8 °C (in dark)	6 months	Monitor at 1 week, 1, 3, and 6 months
Working standard solution	Ambient temperature in light	2 weeks	Monitor at 24 h, 48 h, 1 week, 2 weeks, and with and without protection from light

higher concentration, known as stock standards, and then diluted prior to use. Therefore, it is important to consider both stock (higher concentration) and working standards (lower concentration) in any validation of the standards.

A typical standard validation protocol should make a comparison between a stored standard (stock and/or working standard) and a freshly prepared set of standards. The comparison is made by treating the stored standards as samples and analysing these samples against the fresh standards. The stored standard amounts can then be expressed as a percentage of the expected amount and criteria set for acceptance on that basis. Typical acceptance criteria would be 100 ± 10% of the expected amount. It is obviously necessary to regularly check the standards against these criteria to determine the expiry date and when they should be replaced. For practical reasons you would typically want to validate for long time periods, especially for stock standard solutions which are time consuming or expensive to create. With this in mind, the protocol which you create for validation of standard stability may have varying times and storage temperature conditions which reflect this. An example of this is given in Table 8.6.

8.3.8 How Do I Validate Sample Stability?

Sample stability, as with standard stability, is important to establish as part of the validation of methods. It is especially important if you intend to store samples prior to analysis or if you suspect the analytes in your sample may be unstable, volatile or potentially adsorb onto your collection vessel and thus be lost prior to analysis.

Validation protocols for sample stability should reflect the intended storage conditions prior and during analysis. Sample storage temperature and sample storage time being two of the most important variables to study. More information on sample collection, storage and preservation can be found in Chapter 1.

8.3.9 How Do I Assess the Robustness and Ruggedness of My GC-MS Methods?

When thinking about a newly developed GC-MS method for quantitative analysis it is important to consider the parameters (factors) which may affect the future performance of the GC-MS method, perhaps in a different environment to the one in which the method was developed and validated. It is appropriate to sub-divide these factors into two broad groups:

- Robustness – factors which you can control.
- Ruggedness – factors which you cannot control.

8.3.9.1 Robustness

When a GC-MS method is created there are a very large number of individual parameters to optimise. These include all of the set-points which a typical GC-MS has, such as: instrument temperatures, gas flows and MS related parameters such as detector voltage. Beyond the instrument itself there are parameters set for sample preparation, such as the choice of solvent, extraction and clean-up. All parameters will have an effect on the sample measurement to a lesser or greater degree. One of the key reasons for having excellent method development, is in order to produce a robust method. When optimising a parameter, as the value changes the response will increase, as shown in the example for inlet splitless time optimisation in Figure 8.8. When selecting the optimal value for the method, it should be taken from the point at which the response plateaus and is away from the 'cliff edge' to produce a robust method. Selecting a value to the left of the line, although it would result in only a small change in the response, would be on the edge of a sharp drop in response should the instrument conditions fluctuate, resulting in a less robust method.

8.3.9.2 Ruggedness

Perhaps harder to define, are the factors which affect the method ruggedness, of which you have no direct control over or perhaps are unaware of. These might include:

- Glassware, with factors such as glass quality affecting variation in volume or contamination.
- Environmental factors such as the temperature and humidity.

Figure 8.8   Optimisation of the inlet splitless time, determination of the optimal value away from the 'cliff edge'. Courtesy of Anthias Consulting Ltd.

- Human factors such as competence and training (although training can be fixed).
- Controls on the equipment, maintenance, and software, of which there is not 100% control except to ensure the equipment is as fully maintained as possible.

8.3.10   Why Methods Fail

All of these factors concerning the robustness and ruggedness are part of why a method may fail in the future. These factors are part of the root causes for method failure which are listed below:

- Incomplete description of methods, with information being missing on vitally important parameters such as the GC inlet liner type or important steps in the sample preparation.
- Variations in sample preparation.
- Variations in standard preparation.
- Variations in the column and GC consumables.
- System suitability criteria is not controlling the method well.
- Partial dissolution of samples, or the sample matrix varies greatly.
- Mis-match between sample and standard preparation.
- Too much flexibility in the method.

- Inconsistent integration: the data should be as high a quality as possible and an equal amount of time should be spent on the data analysis method as the acquisition method development.
- Complexity of the method, many steps to complete, or trace analysis in very complex sample matrices with analysis at the LOQ.

To avoid failures in quantitative analysis there are two major ways to assess the robustness and ruggedness of the method.

8.3.11 Design of Experiments to Study the Robustness and Ruggedness

This approach looks to systematically identify the major causes of variation in the method. This exercise is undertaken in a series of steps:

1. Conduct a brain storming session to identify all parameters which might affect the method either as factors you can control or factors which you cannot control.
   - This can be facilitated through the use of techniques such as a cause and effect diagram (Fishbone or Ishikawa) in which the causes are sub-divided into a number of generic causes (method, equipment, people, materials, measurement and environment) as a series of branches from the main "backbone".
2. Repeatedly ask why this is causing an effect to further sub-divide the problem.
   - This process will hopefully result in a complete description of all relevant factors.
3. For each factor, assess how important this factor is to the success of the method.
   - Once again there are several systematic approaches to this, one of which is to score the probability and severity of the risk to the method to determine its importance.
4. For the most important factors, use design experiments to test the effect on the method by varying the factor.
   - For parameters which you control this is relatively straight forward. For factors which you do not control it is just a case of monitoring the parameter when you run the experiment to assess its value during the experiment.

5. Conduct the experiments and analyse the effects.
   - The experimental design can be univariate (in which one parameter at a time is varied, whilst others remain static) or multivariate (multiple parameters are varied at the same time). Multivariate analysis is complex, but a lot quicker to conduct, a full description is beyond the scope of this book but further reading references are provided in the further reading section.

Further Reading

1. P. Bedson, *Guidelines for Achieving High Accuracy in Isotope Dilution Mass Spectrometry (IDMS)*, Royal Society of Chemistry, Cambridge, 2002, pp. 1–34.
2. J. Miller and J. C. Miller, *Statistics and Chemometrics for Analytical Chemistry*, Pearson Education Canada, 6th edn, 2010.
3. AMC Technical Briefs, https://www.rsc.org/Membership/Networking/Interest-Groups/Analytical/AMC/TechnicalBriefs.asp, Accessed June 2019.

Reference

1. Empirical Rule, https://commons.wikimedia.org/wiki/File:Empirical_Rule.PNG, Accessed June 2019.

9 How Do I Maintain My GC-MS?

9.1 Do I Need to Perform Maintenance?

Yes, it is important to perform maintenance on your instrument on a regular basis. For example, you would need to clean the inlet, columns and detectors when analysing samples with high concentrations, complex or high molecular weight compounds which will contaminate them. Maybe the old column you have installed is damaged and is not performing the required chromatographic separation identified through system suitability checks (see Chapter 6)? Hence, you may need a new one. Or maybe the high temperatures used on the inlet may have hardened the septum and liner o-rings, which are now leaking.

Planned maintenance is always better than unplanned trouble-shooting, which will take longer and may result in samples needing to be re-analysed as the instrument performance was not good enough. Overall proper maintenance schedules result in less instrument down time, higher sample throughput and better analysis results.

There are several types of maintenance which are differentiated by the nature of the tasks:

- Preventive maintenance: to maintain a certain level of service on the equipment, programming the interventions of their vulnerabilities in the most opportune time. It used to be a systematic characteristic, that is, the equipment was inspected even if it has not given any symptoms of having a problem.

Gas Chromatography-Mass Spectrometry: How Do I Get the Best Results?
By Diane Turner, Mathias Schäfer, Steven Lancaster, Imran Janmohamed, Anthony Gachanja and Jason Creasey
© Diane C. Turner, Mathias Schäfer, Steven Lancaster, Imran Janmohamed, Anthony Gachanja and Jason Creasey 2020
Published by the Royal Society of Chemistry, www.rsc.org

- Zero hours maintenance (overhaul): the set of tasks for which the goal is to review the equipment at scheduled intervals before any failure appears, or when the reliability of the equipment has decreased considerably so it is risky to make forecasts of production capacity. This review is based on reducing the equipment to zero hours of operation, that is, as if the equipment were new. These reviews will replace or repair all items subject to wear. The aim is to ensure, with a high probability, a good working instrument, fixed in advance of any issues.
- Periodic maintenance (time based maintenance (TBM)): the basic maintenance of equipment made by the users of it. It consists of a series of elementary tasks (data collection, visual inspections, cleaning, *etc.*) for which no extensive training is necessary, but only a brief maintenance training course. This type of maintenance is based on total productive maintenance (TPM).

In this chapter, we will review sections of the GC-MS which require regular preventative maintenance.

9.2 What Should I Do Before I Perform Maintenance?

Before performing any maintenance, the instrument needs to be in a condition to be worked on and also benchmarked. The three principal areas that need to be considered are:

- Benchmarking
- Temperatures
- Gases

9.2.1 Benchmarking

Before taking apart an instrument or performing any maintenance, it is useful to benchmark its performance to ensure that afterwards the instrument is working as well as, if not better than before performing maintenance. Areas to consider, perform and record are:

- What is the vacuum?
- What are the air and water levels?
- What are the tune parameters?

- What does a system blank look like – any peaks?
- Where is the baseline level?
- What does the chromatogram of a standard look like – system suitability checks?

Other ideas for benchmarking can be found later in this chapter in Section 9.5.2.

9.2.2 Temperatures

Depending on the part of the instrument to be maintained, the temperature zones need to be turned off or reduced on the GC-MS:

- Heated autosampler (if applicable), inlet(s), GC detectors and MS transfer line turned off.
- MS ion source and mass analyser temperatures turned off – these may be automatic during venting.
- GC oven cooled to ambient temperature, for example 30 °C – this actively cools the base of the inlet(s) and GC detectors plus the end of the transfer line, enabling them to cool more rapidly. Simply switching off the oven keeps the heat inside!

You should allow at least 20 min before working on it to enable it to cool down. While waiting, gather the tools and consumables needed, see Section 9.3.

DO NOT turn off the carrier gas when cooling!

9.2.3 Gases

Once the instrument components are cool, if working on an autosampler that uses carrier gas, the inlet, column, GC detectors or gas supply, the carrier gas must be turned off JUST BEFORE they are taken apart. Leaving the gas on as long as possible keeps the instrument purged, meaning it will be operational faster after maintenance.

9.2.4 Electricity

In some cases, the power supply to the instrument needs to be turned off, for example when moving the autosampler. At times when the instrument is not working properly, a full power shutdown may be needed, however, care must be taken to follow suitable procedures defined by the manufacturer and ensure that the column is cool first.

9.3 What Tools Do I Need to Perform Maintenance?

It is important to wear suitable personal protection equipment when performing any maintenance on the instrument. This is to prevent any contamination on the instrument from hands and fingers, but at the same time preventing any potentially harmful chemicals transferring onto the body. It is recommended to wash hands and avoid the use of hand creams or perfumes before undertaking any maintenance to prevent contamination and strong odours lingering inside sensitive instruments.

All tools need to be clean and free of grease or other contaminants. They should be stored in sealed contaminant-free boxes when not in use, away from any chemicals.

Use clean and filtered compressed air or nitrogen for any purging of instrument components and lint-free wipes to eliminate particles inside the vacuum manifold and on sealing surfaces. Gloves should be powder-free and if working with any parts that could be damaged with static, for example MS components, gloves should be lint-free.

All parts of the instrument in open manifolds or exposed parts, when you are not working on them, should be covered with aluminium foil to prevent any residual contamination build-up. Where possible, close instrument doors, for example on the MS.

Most manufacturers provide tools with the instrument to ensure that correct items are used for the job and reduce the likelihood of damage to the instrument, always ensure that these are used! The following items are useful:

- Tweezers or long nose pliers
- Longneck Phillips head screwdriver
- Longneck flat head screwdriver
- 3/16", 5/16", 1/8" wrenches
- Allen key wrenches in various sizes
- Wafer for column trimming
- Gloves
- Cotton buds with wooden sticks (not plastic)
- Lint-free cloths
- Sandpaper
- Microgrit/aluminium oxide/autosol
- Solvents: DCM, acetone and methanol
- Four beakers
- Sonicator
- Electronic leak detector
- Electronic flow meter
- Keyboard air duster spray.

9.4   Where Is Maintenance Required and How Do I Perform It?

A schematic diagram of a typical GC-MS instrument, is shown in Figure 9.1. It is recommended to draw up a plan of your instrument configuration, identify the components and from there the maintenance that is required. A typical GC-MS needs regular maintenance in the following areas.

9.4.1   Sample Vials, Caps and Septa

Suitable vials and either crimp or screw caps are needed for the samples. With crimp caps ensure the cap is level and not at an angle, and that it does not spin once crimped. With screw caps do not over-tighten and damage the threads. If the sample volume is very small, special inserts can be introduced into the vial to raise the level of sample in the vial, the autosampler syringe needle penetration depth may need adjusting. Appropriate caps with the correct septa need to be used for the analysis, as certain solvents can cause the deterioration of some septum types. To prevent the contamination of septa from plasticisers, they must not be stored in plastic bags! In highly sensitive analysis, vials and caps can be cleaned with pure methanol approximately three times, dried with clean gas and placed in an oven at 70 °C to remove any contamination or simply thermally conditioned at this

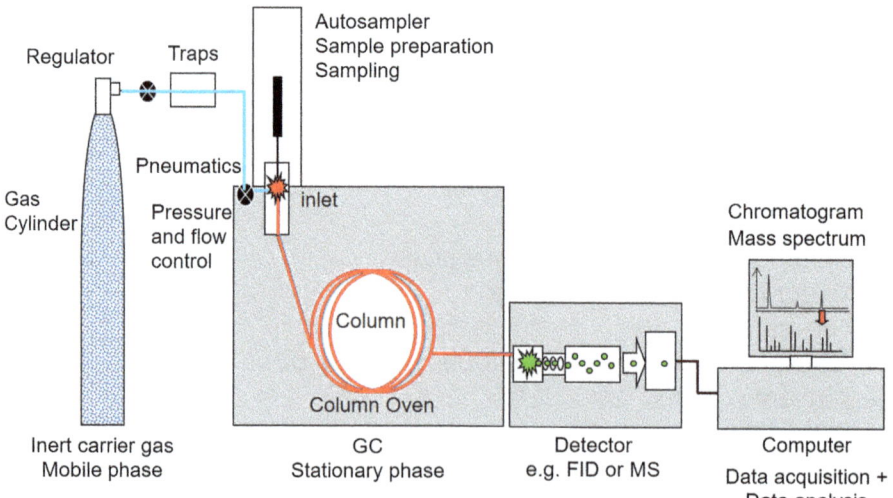

Figure 9.1   A schematic diagram of a GC-MS instrument which can be important for reviewing and understanding the places in which maintenance is required. Courtesy of Anthias Consulting Ltd.

temperature. Conditioned vials and caps should be stored in glass jars with cardboard-lined lids before use.

If the samples are to be kept, the cap and septum should be replaced as soon as possible after injection to prevent sample evaporation and contamination, therefore screw-top vials can be more appropriate for this. Samples should then be suitably stored ready for re-injection.

The autosampler should be regularly cleared of used vials, which should be disposed of appropriately.

9.4.2 Maintenance of the Sampling Device or Autosampler

9.4.2.1 *Liquid Autosampler*

9.4.2.1.1 Syringes

Syringes require proper care to prolong use. Appropriate cleaning and handling of each syringe will help ensure correct analytical performance and a long syringe life.

- Rinse syringes and clean their plungers before use to maximize syringe lifetime.
- Rinse the syringe 5–8 times between injections to minimize sample carryover.
- Pump sample in and out of the syringe at least five times to remove any air bubbles, and for maximum reproducibility and accuracy. If the sample contains particles do not perform this step, as needle blockage may occur.
- Perform a visual check of the needle and plunger daily.
- Remove the syringe and manually clean it weekly, checking for:
  o A worn and leaking plunger – replace; a partially blocked needle – replace needle or syringe.
- A 26-gauge syringe can be used for on-column injections into a 0.53 mm internal diameter (i.d.) column. Always check that on-column syringe needles fit inside the capillary column before installing the column and syringe.

When cleaning your syringe, it is best to use solvents that effectively dissolve the sample you are working with. Try to avoid cleaning agents that are alkaline, contain phosphates, or are strongly acidic. Non-polar solvents can, over time, cause the plunger to stick – consider using methanol as the second wash. PTFE tipped plungers can also help with sticky samples. Always ensure plenty of washes are performed.

9.4.2.1.2   Wash and Waste Solvents

It is important to change dirty/empty wash solvents.

- Daily: fill wash vials and empty waste vials.
- Weekly: clean wash and waste vials, replace septa or clean vial needle guides.

9.4.2.1.3   General Autosampler

Monthly: clean needle guide, any moving parts, wash and waste vial stands, and sample trays to prevent dust and grease build-up.

9.4.2.2   *Thermal Desorption Instruments*

The Thermal Desorption (TD) unit can suffer from leaks, contamination and loss of analytes. The unit may also suffer from carryover, appropriate temperatures and flows need to be considered when creating the method to avoid these problems. The TD unit may employ internal leak checking processes to prevent loss of sample, stop the analyses and report an error.

Regular maintenance on the instrument includes:

- TD tube conditioning and re-conditioning;
- replacement and cleaning of TD tube end caps and o-rings;
- re-conditioning and replacement of the cold trap;
- replacement of o-rings, transfer line and fittings within the instrument flow path.

The sorbent used in the TD tubes and cold trap will need replacing, they can be re-packed after a defined number of method (heating) cycles as the packing material will break down and deteriorate in performance. The frequency is dependent on the sorbent type, for example approximately 100 cycles for Tenax.

9.4.2.3   *Pyrolysers*

There are several different pyrolysers types with different configurations, with problems including leaks, the sample size being incorrect or non-homogenous and incorrect pyrolysis temperatures being used. Depending on the type, maintenance could include:

- Cleaning the quartz cell/chamber/tube.
- Replacement of the septa and/or o-rings.

- Broken filaments: replace and re-calibrate.
- Blocked, broken, contaminated transfer line or it not heating – clean or replace.

9.4.2.4 Headspace Instruments

Syringe-based or valve-based headspace instruments require slightly different maintenance. However, they can all suffer from incorrect matrix modification, inconsistencies between samples as they were not 'normalised', leaks on the vials and instrument, contamination and carryover owing to incorrect temperatures. Maintenance includes:

- Running blanks after high concentration samples.
- Regularly leak check fittings.
- Replacing the syringe, syringe plunger or o-rings if leaking or contaminated (check temperature).
- Replace the transfer line if not heating, contaminated or leaking.
- Replace fittings and tubing if leaking or contaminated.
- Cleaning/dusting moving parts.

9.4.2.5 Purge and Trap

Purge and trap maintenance is very similar to that on TD and headspace systems. This includes:

- Regularly flush lines with hot water, particularly after high concentration or foaming samples.
- Check for leaks daily.
- Replace cracked/leaking/contaminated glassware.
- Clean or replace stuck/leaking/contaminated valves, tubing and fittings.
- Cleaning or replace a blocked or contaminated transfer line.
- Replace the cold trap and water management trap.

9.4.2.6 Solid Phase Extraction, Solid Phase Microextraction and Similar Techniques (Stir Bar Sorptive Extraction)

Mostly consumables are used on these instruments. During method development, it should be determined how many samples the phase can be used to extract. Maintenance or use should include:

- Conditioning of the phase before use.
- Re-conditioning after use, particularly for high concentration samples, this could just be a blank run.
- Replacement of the phase when damaged.

9.4.3   Maintenance on the Gas Chromatograph

The GC maintenance can be divided into multiple modules: gases and plumbing, inlet, column and column oven, plus any GC detectors.

9.4.3.1   Gas Maintenance

High purity carrier gases are essential for consistent and accurate analyses. Proper gas maintenance is key to achieving this goal. Gas purifiers, regulators, leak detectors and flowmeters are all designed to prevent column damage, reduce MS maintenance and improve the quality and consistency of your separations.

9.4.3.1.1   Cylinders and Generators

Cylinders store gases above atmospheric pressure at around 3000 psi. Safe lab practice should be used to transport, store and use cylinders.

- Ensure safety caps are in place over the cylinder gas valve.
- Always chain/strap cylinders to walls when in use or storage.
- Check the contents weekly and record – helps to identify leaks as the volume left should slowly reduce, not suddenly drop.
- Replace the cylinder when it reaches 30 psi. DO NOT EMPTY as impurities like moisture, hydrocarbons and small particulates from the cylinder can enter the gas supply.
- Cylinder changes: use a manifold, multiple-stage regulator or toggle valve to keep the lines pressurised to minimise contamination into the system, saturation of the traps to minimise and loss of expensive gas upon changing.
- Gas generators need servicing every 1–2 years (in particular, filters need replacing), check with the manufacturer.

9.4.3.1.2   Regulators

A gas regulator controls the pressure of gas into the supply lines and has two gauges. Gauge 1 indicates pressure left in the cylinder and gauge 2 indicates the pressure being delivered. It is important to ensure there is enough pressure being delivered to the GC(s) to give the required flows + 10 psi.

- Check the regulator for leaks when changing the cylinder.
- Overhaul regulators every 2–5 years and replace every 5–10 years when required. The frequency for corrosive gases is every 1–2 and 2–4 years respectively.

An important consideration is that there are different regulators for different gases: air or hydrogen or helium/nitrogen/argon or oxygen – so use the correct one! (check with the gas supplier if in doubt).

9.4.3.1.3 Supply Tubing and Joints

Gas supply tubing should be made from either analytical grade stainless steel (SS) or copper. SS is recommended for reactive hydrogen and plastic should never be used as it is porous and a source of plasticisers.

Fittings such as joints, tees, branches, valves, and so forth may start to leak and require checking every 6–12 months, as well as when replaced to prevent any loss of gas or ingress of contamination.

It is recommended to use toggle or shut-off valves between instruments to isolate them if required without switching off the gas supply.

9.4.3.1.4 Traps and Filters

Clean, dry and pure carrier and detector gases are required in GC-MS systems to ensure optimal performance and to reduce instrument maintenance. Traps should be installed just before the GC-MS to remove:

- Oxygen which can damage column phases at temperatures greater than 60 °C and will reduce the performance of electron capture detectors (ECDs).
- Water can also damage some column phases and will reduce the performance of the MS.
- Hydrocarbons cause background contamination in flame ionisation detectors (FIDs) and mass spectrometers.

Traps are installed (in the order from cylinder to instrument): hydrocarbon trap (activated charcoal), moisture trap (molecular sieve) and oxygen trap (aluminium oxide).

They require regular replacement depending on the purity of the gas, volume of gas used and any leaks, on the manufacturer's recommendation or at least every 1–2 years. Indicating traps are very useful, as they only need to be replaced when 75% saturated.

9.4.3.1.5   Electronic Pressure or Flow Controllers

The internal pressure and flow controllers (electronic pressure controllers (EPCs), automatic flow controllers (AFCS)) that control the flow through the column or detector may at times be damaged giving wrong readings. Certain instruments have these devices calibrated or leak tested to ensure there is no loss or degradation. Some controllers contain o-rings that can be replaced if leaking.

9.4.3.2   Inlet Maintenance

There are various parts to the GC instrument which require maintenance, of which the most important part is the inlet. It is considered that 90% of all chromatography problems can be associated with this module, especially with dirty samples, as it is the site of sample injection and evaporation.

9.4.3.2.1   Septum and Septum Cap

The septum enables the introduction of the sample into the inlet without pressure loss and seals around the needle as the sample is injected to exclude air and contaminants.

When installing the septa, you should be careful not to over-tighten the nut as it will compress the septum causing leaks and may damage the needle when it is inserted. Under-tightening the nut may cause inlet pressure failures and leaks of the sample when injecting, providing an additional split flow.

Most septa are purchased pre-conditioned, however if unsure, they may be conditioned by holding at 50 °C above the expected operating temperature for an hour. Cycle through the oven temperature program to be used at least twice after conditioning to clear any residues from the system.

There are several types of septa available, but key features to consider are:

- Maximum operating temperature: most septa are fine up to 350 °C, above this a high-temperature septum is required, for example, BTO silicone.
- Size: use of the recommended depth and diameter of the septum is necessary to achieve a good fit and prevent leaks.
- Material: this generally dictates the temperature profile, level of bleed and the number of injections before replacement is needed, it is usually silicone of various grades. BTO (bleed and temperature optimised) silicone can be used at higher temperatures but

Figure 9.2 Image of an inlet that severely requires maintenance. Courtesy of Anthias Consulting Ltd.

is a harder material enabling fewer injections to be made and septum coring to be more pronounced.

- Pierced: non or pre-pierced septa. Non-pierced septa can easily be pierced once installed with a syringe!
- Coating: most septa are plasma coated to be non-stick. PTFE or Teflon coated can be useful for inlet septa not pierced with a syringe needle.

The key factors in how frequently the septum needs to be replaced are:

- Number of injections made through the septum.
- The needle style of the syringe – a cone-tipped needle will greatly reduce coring (particles of the septum can drop into the liner and be analysed, see Figure 9.2) and prolong septum life.
- Solvent type used – check with the manufacturer which solvents the septum is compatible with.

For high throughput analysis, the septum may be changed every day. When the septum is not pierced it should be replaced every 6–12 months when it hardens.

The septum cap will also become dirty and contaminated and will require a monthly clean with solvent (preferably methanol or a final clean with methanol) and a clean lint-free cloth or wooden cotton bud, particularly in the needle guide.

9.4.3.2.2 Liner and Liner Seal

The liner is the centre piece of the inlet, into which the sample is injected. It is very important to select the correct liner for the inlet, injection technique and analytes, as discussed in Chapter 2. If lots of dirty samples are being analysed the liner will quickly become contaminated

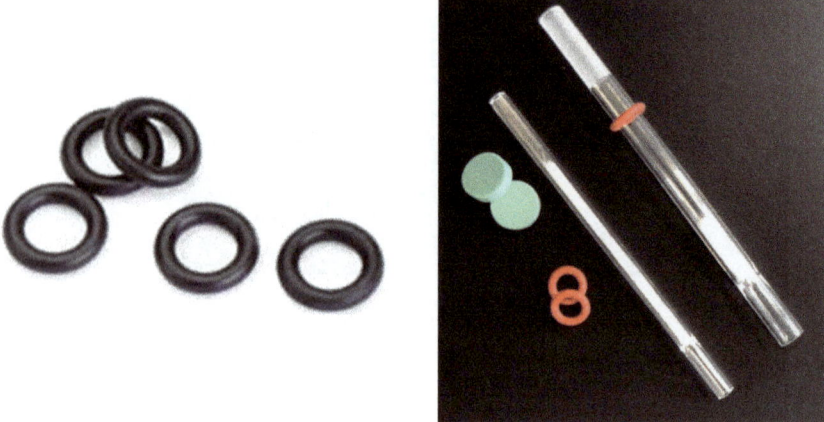

Figure 9.3   Inlet liner o-ring seal and inlet liners. Courtesy of Anthias Consulting Ltd.

and will require far more frequent replacement than with clean samples. The polarity and concentration of the analytes are also a consideration, as very active low concentration analytes, even in a clean sample, could show signs of activity through tailing and even disappearing after one batch of samples. Therefore, how frequently the liner is replaced – daily, weekly or monthly is dependent on the application.

When removing a dirty liner for replacement, always examine it.

- Dirt on the outside means the capacity has been exceeded and now the inlet body will need cleaning and the column trimming. More method development is also required to prevent re-occurrence – see Chapter 2.
- Wool in the centre of the liner with dirt at the bottom of the wool may mean that dirt has also progressed onto the column – again trim it.
- Dirt in the taper, means the inlet body will need cleaning, the inlet seal (if present) will need replacing and the column trimming. A liner with wool near the bottom is better for this application.
- Dirt on top of the wool, indicates that all of the dirt is contained in the liner and after replacement no further maintenance is required.

The liner seal or o-ring separates the carrier gas entering the liner from the split line and can also seal the top of the inlet. Depending on the inlet type, it can be a graphite seal (usually on some PTV inlets) or fluorocarbon/viton/silicone o-ring (see Figure 9.3). Over time, temperature and liner replacement can damage the liner seal or o-ring so that it becomes hard, cracked and squashed losing its ability to seal the liner within the inlet body and give consistent results.

Each time the liner is replaced the seal or o-ring should be examined for damage and replaced if necessary.

9.4.3.2.3 Inlet Body

The inlet body can become contaminated with high matrix samples and should be cleaned monthly for high-throughput analysis with solvents (acetone then methanol) as part of its maintenance. Make sure to remove the column first and always do a final clean with methanol.

9.4.3.2.4 Ferrules

There are various types of ferrules available and the use of the correct one is important to prevent any leaks on the instrument. The purpose of the ferrule is to provide a seal between unions, valves, transfer lines and devices. The common types of ferrules, along with their advantages and disadvantages, are:

- *Graphite ferrules*: these are soft, easy to remove and reusable, do not start to leak when the temperature is changed as with PTV or cool on-column inlets, but they are porous to oxygen and hence not suitable to for MS or ECDs. They should not be over-tightened on installation, as this compresses the graphite. They can also flake or fragment causing particles to remain in the injector or detector, therefore the column should always be trimmed after ferrule installation! They can be reused but check the condition first.
- *Graphitised vespel ferrules* (85% vespel 15% graphite): these are hard and non-porous ferrules good for MS transfer lines and ECDs but can leak when the temperature is changed. You will need to ensure that you re-tighten these ferrules after initial installation and temperature cycling then regularly check thereafter – the most likely place to leak in GC-MS is the end of the transfer line in which this ferrule is installed! They can be reused but check the condition first.
- *SilTite^{TM} or SilFlow® ferrules*: these are soft metal ferrules which are totally non-porous and clean (good for MS). They can be used in selected systems and use a unique nut. Both nut and ferrule expand and contract together with temperature changes, resulting in a leak free connection. The ferrule seals to the column polyimide coating and provides a permanent column connection especially in high pressure systems prone to leaks, but this means the column must be trimmed to remove it. If the column is to be used again without trimming, storage caps can be used to ensure the end of the column stays in good condition when not in use.

9.4.3.2.5   Inlet Seal

Some instruments have a replaceable inlet seal positioned at the base of the inlet below the liner which comes into contact with the sample. Sample residue can accumulate on the seal, which can be easily replaced when it gets dirty to prevent adsorption of active compounds.

The procedure to maintain this part, is to observe its condition from the top of the inlet each time the liner is changed. Small amounts of dirt or particles can be removed with a cotton bud and methanol, however the column must be removed first! If it is very dirty and needs to be replaced, open the base of the inlet from inside the GC oven and replace it with a new seal and washer. Never reinstall a used seal, as the groove created from the previous installation may leak.

There are various types of seals – stainless steel, gold plated and inert Siltek™ treated. Gold plated is the most common, known as a 'gold seal'. Siltek™ treated seals are used for the analysis of very active compounds to reduce breakdown and adsorption.

9.4.3.2.6   Split Vent Line and Trap

The split vent line runs from the side of the inlet to the split line trap. High molecular weight (MW) compounds exiting the inlet through the split line to waste can condense inside the tube before they reach the trap. Over time (years depending on the number of samples analysed, the split ratio and how dirty the samples are) the line will become clogged then blocked and will need replacing. The symptom is – no split flow! On many instruments, the split line is 1/8″ tubing rather than 1/16″ so that it doesn't need such frequent replacement.

The split vent trap is used to trap the more volatile split exit effluent including solvent so that it does not enter the air in the laboratory. This will need replacing every 6 months to 2 years depending on the split ratio used (higher split ratios send more sample to waste) and the number of samples. This is very easy to perform and does not require the split flow to be turned off, just reduced. Split vent traps may be installed with 1/16″ nuts or with a finger tight holder as shown in Figure 9.4. Ensure the trap is installed the correct way around with the flow from the inlet to the exhaust.

9.4.3.3   Column Maintenance

9.4.3.3.1   Column Care and Storage

Capillary columns are durable and robust, but they must be handled with care to avoid outside contamination and damage.

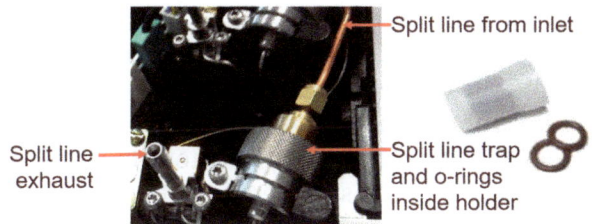

Split line from inlet

Split line exhaust

Split line trap and o-rings inside holder

Figure 9.4 A schematic of the split line, split line trap and exhaust on an Agilent GC. Courtesy of Anthias Consulting Ltd.

It is essential to:

- Avoid scratching the polyimide coating by storing it in the box.
- Protect it from air and contamination when not installed by sealing it at both ends with a septum.
- Avoid injecting dirty/high MW samples, inorganic acids and bases.
- Ensure the instrument is leak tight with low air and water levels.
- Avoid bending the column unnecessarily or the column touching the walls of the GC oven.
- High-boiling, non-volatile compounds may accumulate at the inlet end of the column affecting the chromatography so that the peak shape deteriorates. The column can be trimmed to remove the contamination (for some applications this is performed routinely when the liner is changed) or, if installed, backflushing may be used to remove it.
- When installed in the instrument but when samples are not running, it is recommended to use a standby method to keep the oven temperature cool (<60 °C but preferably 30 °C) and always maintain column flow at 0.5 to 1 mL min^{-1}, especially when connected to an MS.

9.4.3.3.2 Column Trimming Ready for Installation

The shape of the column ends is critical to the performance of the method. They should be square and accurate to minimise the active surface area and provide smooth flows.

- Wash your hands. Some people prefer to wear gloves but it is difficult to determine the amount of pressure applied when trimming.
- Place the nut on the column.

- Place the ferrule on the column the correct way around for the nut (some require the pointed end of the ferrule to point into the nut, others it is the opposite – check first).
- Place the column onto your finger.
- Find the smooth edge of the wafer (there is a smooth edge for fused silica columns and a serrated edge for metal columns).
- Place the far edge of the wafer (not the middle) onto the column ensuring it is perpendicular to the column and at an angle of 45° (wafer to column) to use the smooth edge.
- Apply gentle pressure and score in one sweep until the end of the far edge (but stop before the edge so that it doesn't damage the column).
- Gently push the column onto your finger away from you (but do not break it completely off).
- Flick the column back towards yourself with your finger, not the wafer, to break the far edge.
- Inspect the column (with a magnifying glass if needed).

If the ferrule is dropped, replace and re-trim the column to avoid any blockages.

9.4.3.3.3 Column Installation Distances

Correct column installation into the GC inlet and detector is very important for good chromatography. The exact distance that the column should be installed into each module is both manufacturer and module type specific and the user needs to review the instrument manual to ensure the correct length for the type of inlet or detector is used. The lengths can vary from 1 mm to several centimetres measured from the tip of the ferrule to the tip of the column. For example, an Agilent 7890 split-spitless inlet has a length 4–6 mm (aim for 4 mm), whilst a Shimadzu GC2010 split/splitless inlet has a correct length of 34 mm. An Agilent MS in the electron ionisation (EI) mode requires 1 mm out of the transfer line, whereas an Agilent FID is 48 mm of column from the tip of the ferrule.

If the column is installed too far into the inlet it is difficult for the analytes to transfer onto the column, resulting in a loss of sensitivity, especially for active compounds. In the detector (including MS) if the column is installed too far in it interferes with the detection and may burn the end of the column; if it is too far out there is a long distance for analytes to travel into the detector resulting in reduced sensitivity or even a loss of less volatile analytes.

9.4.3.3.4 Column Conditioning and Restoration

Once correctly installed, it is important to condition the GC column in order to remove harmful vapours (oxygen and water), any residual contamination and column bleed. The process is to:

- Install the column into the inlet, but do not connect to the detector, leave the free end in the column oven so that bleed does not contaminate the MS (Note: this is not suitable if your carrier gas is hydrogen. In that case, connect the column to the detector too).
- Apply carrier gas at a typical flow rate for your column or the application (or twice as high to purge quickly) and allow the carrier gas to flush the column at an ambient temperature for 10 to 30 minutes to remove any air and in particular oxygen from the column before heating.
- Increase the oven temperature to 105 °C ramping at 10 °C min^{-1} and purge the column for a further 15 to 30 minutes to eliminate any residual moisture.
- Finally, programme the temperature to rise by 10°C min^{-1} to the maximum temperature of the method or the column's maximum isothermal temperature with a 60 to 240 minutes hold to remove excess column bleed and any contamination.
- Note: purge times are dependent on the type and amount of stationary phase in the column. Longer columns and thicker or polar stationary phases will take longer to purge and longer to stabilise the column bleed.

Once complete, cool the oven and install the column into the detector (if not already done so).

If the column becomes contaminated it can be reconditioned as above, but with reversed flow so that the end that was installed into the detector is now installed into the inlet. If this does not work and the column is long enough, 0.5–1 m can be removed. If that does not work rinse with solvent following the manufacturer's instructions, for example rinsing with a 9:1 mixture of dichloromethane : methanol. If that does not work the column is dead – replace with a new one!

9.4.3.3.5 Guard Columns

In some cases, it is necessary to have a sacrificial "guard" column to protect the analytical column from contamination. Guard columns are typically uncoated deactivated silica and do not have any stationary phase but are connected *via* a union, tee- or press-fit connection. Depending on its use, it can also be called pre-column or restrictor. A guard column benefits from the following:

- Protects the column from liquid solvents and non-volatiles as these condense on the guard column.
- Prevents damage to and trimming of the analytical column, avoiding the effects of retention time shifts. For example, 60 cm removed from a guard column will reduce the retention times by only 3 s, but without a guard column, there could be 2.5 times greater retention losses.
- Helps to condense and focus solvents such as methanol when using a non-polar stationary phase or hexane in a polar stationary phase for the solvent effect in splitless injections (match the polarity to that of the solvent).
- Is a useful aid for on-column injection, enabling larger volumes to be injected and narrower i.d. analytical columns to be used.

Some columns already have integrated guards: for example, a 30 m column with a 5 m integrated "segment" and the user should ensure it is installed correctly in the correct position, in that the guard column is on the inlet end. The guard column length should also be included when entering the column dimensions into the instrument acquisition software, as this will affect pressures for constant flow methods as there is a greater column length.

9.4.3.4 GC Detector Maintenance

The exact maintenance required depends on the detector.

Routine maintenance of a flame ionisation detector (FID) should include:

- Ignite FID – monitor background and noise levels, check baseline is stable.
- Check pressures/flows.
- Clean or replace jet.
- Inspect igniter assembly, replace if necessary.
- Clean collector assembly and inspect connections.
- Remove, trim and reinstall column.

As an electron capture detector (ECD or μECD) has a radioactive source, it is a sealed unit and usually (manufacturer dependent) cannot be opened or dismantled. Routine preventative maintenance is restricted to thermal cleaning (baking-out) of the detector:

- Cool ECD and remove capillary column from detector.
- Cap detector inlet.
- Maintain make-up flow through detector and raise detector temperature to 350–370 °C.
- Monitor output of detector to assess decontamination.
- When stabilised/back to normal level, reduce temperature and re-install column.

The ECD must be monitored to ensure that radioactive material does not exit the detector and therefore routinely undergoes a wipe test, the frequency of which is dependent on the country in which the instrument is located.

- Several areas of the detector are "wiped".
- The swabs are analysed for residual radioactivity.

9.4.4 Maintenance of the Mass Spectrometer

Before a GC-MS instrument is used, at a minimum the status should be checked with respect to the vacuum and air/water levels. Failure to identify issues with these could result in a large problem with the MS system if it is used without resolving these first.

9.4.4.1 *Performing an Air and Water Check*

The air/water check on the instrument needs to be performed every day to observe the presence of air and moisture levels in the entire GC-MS instrument, not just the MS. Higher levels can reduce the performance of the GC-MS and if large enough can affect the vacuum system. Different instruments have different approaches to monitoring air levels, some look at the ratio between the nitrogen and water and others look at the relative ratio between each of these and the *m/z* 69 ion present in the EI tune compound (PFTBA). The peak width, absolute abundance or relative abundance of the ion *m/z* 28 for nitrogen is monitored. Water levels are observed using the ion *m/z* 18.

Figure 9.5 shows an example of an air/water report from an Agilent GC-MS. Water is measured as 3.16% relative to *m/z* 69 and nitrogen at 1.49%. For optimal performance nitrogen should be less than 5% and water less than 10% for this instrument, with working values of nitrogen being less than 10% and water less than 20%. Therefore, this instrument can go on to be tuned or used for sample analysis with a

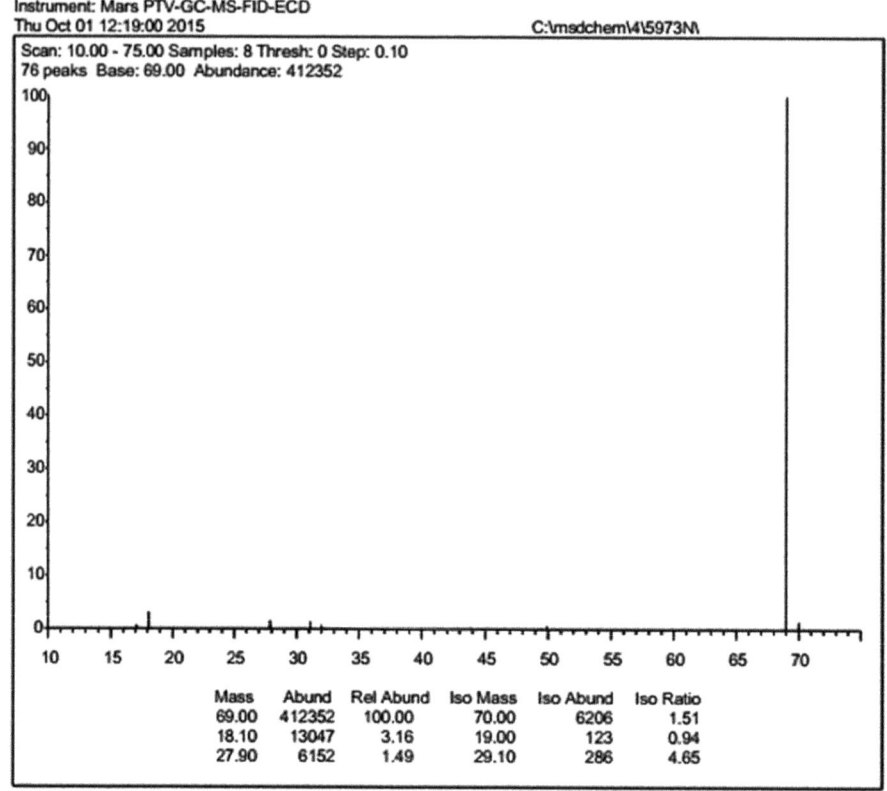

Figure 9.5 An Agilent EI Air/Water report obtained using MSD ChemStation software. © Agilent Technologies, Inc. Reproduced with Permission, Courtesy of Agilent Technologies, Inc. The RSC accepts no liability for the accuracy or reproducibility of screenshots.

high performance. It is important to research the criteria used by the manufacturer of your instrument.

9.4.4.2   Performing MS Tuning, Optimisation and Calibration

Depending on the manufacturer, the instrument tuning and calibration can be manual or automated with tuning and calibration taking place in one action or separately. The tuning and calibration optimises performance by adjusting parameters to improve mass resolution and sensitivity. It usually optimises the voltages to ionise the analytes, accelerate and focus the ions, determines the mass analyser gains and offsets to obtain the correct peak width for mass resolution, and sets the detector voltage to achieve a certain abundance of a certain m/z of the tuning compound. It finally calibrates the mass axis for the correct mass assignment.

The instrument tune report contains a valuable amount of information about the instrument state. It can indicate leaks, a good or poor vacuum, chemical contamination, a dirty ion source, an exhausted electron multiplier and malfunctioning heated zones.

The frequency of the tune depends on the stability of the instrument, it could be daily or weekly or monthly, and is instrument specific. There are also multiple different types of tuning methods from low mass autotunes for volatile analytes, autotunes to obtain the best sensitivity, tunes to force the ion ratios to be the same as for a magnetic sector instrument, as well as different tunes for EI, positive chemical ionisation (PCI) and negative chemical ionisation (NCI) techniques. Therefore, select the one most appropriate for your application. Figure 9.6 shows an example tune report from an Agilent GC-MS EI autotune.

9.4.4.2.1 Electron Ionisation Tuning Compound – PFTBA

Perfluorotributylamine (FC43) (PFTBA) is used in most GC-MS instruments with EI as the tuning compound. It has known properties, is volatile, stable and fragments over a wide mass range with the key tuning ions being m/z 69, 219 and 502.

The tuning compound is held in a vial behind a valve and the level should be checked every 3 months. When topping up or replacing, the vial should not be over-filled. After reinstalling, the air must be purged out otherwise the filament or ion detector could become damaged.

9.4.4.2.2 Chemical Ionisation Tuning Compound – PFDTD

Perfluoro-5,8-dimethyl-3,6,9-trioxydodecane (PFDTD) is commonly used as the tuning compound for chemical ionisation (CI). It is a pure gas, held in a separate vial to the PFTBA. PFTBA tuning usually must be performed first. PFDTD tuning is performed infrequently, as it is known to dirty the ion source.

9.4.4.3 Vacuum System

9.4.4.3.1 Breaking the Vacuum

The vacuum system should always be turned off in a controlled manner:

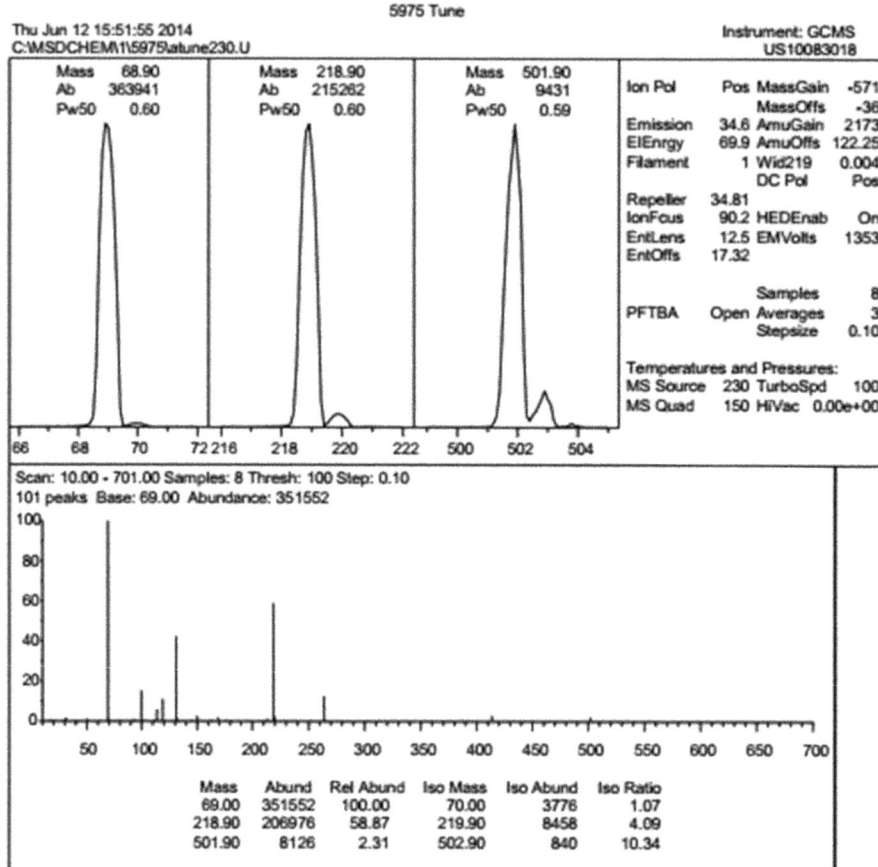

Figure 9.6   An Agilent EI Autotune report obtained using MSD ChemStation software. © Agilent Technologies, Inc. Reproduced with Permission, Courtesy of Agilent Technologies, Inc. The RSC accepts no liability for the
accuracy or reproducibility of screenshots.

- Turbos should gradually slow from 90 or 60 000 rpm.
- Diffusion pumps need to cool to prevent suck-back of oil through
  the MS.
- The mechanical (back-up) pump should operate until the turbo/
  diffusion pump has stopped.

- The MS ion source and mass analyser should cool to less than 100 °C before air is allowed into the system, otherwise air molecules can pit the metal surfaces.
- Remember to cool down the GC temperatures too, for example, at a minimum, the transfer line and GC oven.
- While waiting collect the tools and other items required to perform the maintenance – gloves MUST always be worn!

Some MS instruments have an automatic shut-down at the click of the button, others require the steps to be carried out manually. Once complete the MS power button can be turned off (if not already) and the mechanical pump turned off (if still not off).

Air should only be allowed into the system once maintenance is ready to begin. Use the vent valve to break the vacuum, turning only until a hissing can be heard indicating that air is being sucked in. Open further once the hissing has stopped and repeat until it hisses no more – do not completely remove the vent valve! The vacuum should not be broken through the transfer line, as small particles from the ferrule could be sucked in.

9.4.4.3.2 Pumping Down the Vacuum

As with breaking the vacuum, pumping down the MS must be carried out in a controlled manner. Depending on the manufacturer it may all be automated, partially automated or completely manual. The exact steps are manufacturer-specific but are usually something such as:

- Ensure the GC is on with column flow of around 1 mL min^{-1}.
- Ensure all openings are closed – work your way through the system: inlet, column, transfer line nut, vent valve, and so forth.
- Wipe any MS seals with a lint-free cloth (if needed, clean with methanol, if absolutely needed lightly lubricate with Apiezon L high vacuum grease).
- Press on the MS door or side plate to get a good seal – it is preferable not to use screws to hold this closed as it may not seal properly, especially on older instruments.
- Turn on the main power switch and the mechanical pump too if it is not powered by the MS.
- The mechanical pump turns on first to take the MS from ambient to low vacuum (hard work!). The gurgles should reduce in less than 30 s otherwise there could be a large leak (did you forget to close a nut somewhere? In which case turn-off the MS and fix it.)

- You should feel the fan is working through the vents on the MS.
- The MS will be booting up, check the display.
- There may be a beep to indicate that a low vacuum (10^{-2}) is achieved or you might see from the vacuum gauge or instructions on the screen: release the side plate or door.
- The high vacuum pump (turbo or diffusion pump) now kicks in to reduce the pressure further and reach a high vacuum.
- If it didn't before, the software should now connect to the instrument or a vacuum reading should be seen.
- Do not heat anything until a good vacuum is established and the high vacuum pump, if a turbo, is up to speed (100%).
- If you have leak and need to vent again it takes a long time to cool!

9.4.4.3.3 Mechanical Pump Maintenance

The mechanical pump reduces the vacuum from ambient to 10^{-1}–10^{-2} Torr; serves as a backing pump to the high vacuum pump; traps un-ionised molecules, analysed ions and any molecules evacuated from system within the oil; and traps oil vapour in an oil-mist trap. The oil in the pump provides the seal that enables the pump to create a vacuum and therefore proper maintenance, including routine oil changes, is critical to obtain a good vacuum.

- Weekly: Check the oil level (should be between the maximum and minimum marks), check for signs of any leaks into the pump tray and top-up where necessary (requires venting to perform this).
- Weekly: Check the oil color, which indicates the degree of thermal ageing and change every 6–12 months or when the oil color is a very light orange.

To change the oil:

- Vent the MS and allow the oil to cool but not go cold (warm oil can be removed much easier).
- The oil contains compounds which were not ionised in the MS along with everything else. If the samples being analysed over the previous 6–12 months contained any toxic compounds then these are now concentrated in the oil. Always wear gloves, safety glasses and a lab coat and change the oil in a fume hood if possible.
- Remove the drain plug and decant the oil into a suitable container.
- Rinse the pump through with a small amount of clean (clear in colour) oil and drain.

- Replace the drain plug and fill with clean oil, suitable for the pump, up to the minimum line.
- Allow to settle then top-up further – DO NOT go more than half-way between the minimum and maximum lines. When the oil heats up it will expand and the maximum line must never be exceeded, otherwise it will damage the pump!
- If the pump is fitted with an oil mist filter, clean or replace this. To avoid this consumable in the future, connect the pump vent to an exhaust or a fume hood to remove oil vapour, toxic solvents, chemicals or flammable carrier gas from the lab air.
- Close the fittings and reconnect the pump, ensuring any traces of oil are removed and that the pump tray is clean (making it easy to see if a leak develops).
- Pump down the MS as described, ensuring that the vacuum is working well and in particular that the mechanical pump does not gurgle too much or for too long compared to normal.
- Dispose of the oil appropriately – as toxic waste if necessary.
- Oil that has reached a brown colour will have damaged the pump and it will require a full service prior to use.

Bench-top GC-MS instruments usually use small mechanical pumps that do not require ballasting. However, larger mechanical pumps on instruments in heavy use in EI require ballasting once a week; MS instruments in CI mode using ammonia require ballasting for 1 hour per day; or mechanical pumps used on particular detectors such as sulphur chemiluminescence detectors (SCDs) require ballasting every 1 to 7 days, depending on use. Ballasting feeds air into the working chamber of the pump *via* the gas ballast valve to avoid condensation and to ensure that any solvents dissolved in the pump oil can escape. When the MS is not operating, switch the ballast valve to the on position and leave for 1 hour, the pump will be louder and may gurgle. Do not leave open otherwise oil will be lost and the pump damaged. Switch back to the closed position and use the GC-MS as normal.

It is wise to seat the mechanical pump on a clean tray which can then be used to check for any oil leaks. There are numerous seals in a mechanical pump which will need replacing over time, identified by a leak.

9.4.4.3.4 Diffusion Pump Maintenance

Diffusion pumps are high vacuum pumps that work by boiling oil. On re-condensing, gases are trapped in the vapours and transported to the pump outlet where they are sucked through to the mechanical pump.

The colour and level of the diffusion pump oil should be checked every time the system is vented or at least every 6 months. If the oil is dark, cloudy or below the minimum mark it should be replaced.

9.4.4.4   Ion Source

As described earlier, the ion source is where the column effluent enters the MS and it is the site of ionisation of neutral molecules by the electrons produced by the filaments. The ion source can become very dirty, how quickly depends on factors such as the level of column bleed, cleanliness of the sample, volatility of the matrix and the number of injections.

The need to perform maintenance of the ion source can be observed by increasing voltages observed in the tune reports to push and focus the ions (repeller/push plate/focusing lenses), the detector voltage, quality of the mass spectra, any loss of sensitivity or increase in the baseline.

Maintenance such ion source cleaning is best left until the last day of the working week so that the system can equilibrate over the weekend. However, the system should be fully checked for leaks before leaving it!

Exactly how to clean the ion source is again manufacturer and instrument dependent, but as an example for an Agilent GC-MS, the steps could be:

- Vent the MS as previously described.
- Lint-free gloves must be worn to remove the ion source.
- The MS door should be closed to prevent contamination of the rest of the system while source cleaning takes place.
- Wear lint-free gloves to take the ion source apart (Figure 9.7) and separate components into those that can be cleaned (100% metal) and those that cannot (anything with ceramics, plastic or wires usually cannot be cleaned – but check the manufacturer's instructions). If possible, remove the delicate filaments first, then when re-assembling put them back on last.
- Physical cleaning of any 100% metal components can then take place.
  - o Disposable gloves should be worn as they will get dirty.
  - o Abrasives such as the green paper shown in Figure 9.7 can scratch metal surfaces so minimise their use.
  - o It is better to use microgrit (aluminium oxide) mixed in methanol or water to remove visible marks with cotton swabs on wooden sticks and to clean other surfaces too.

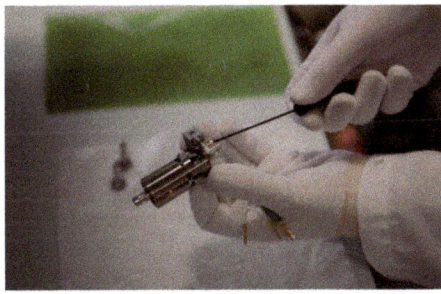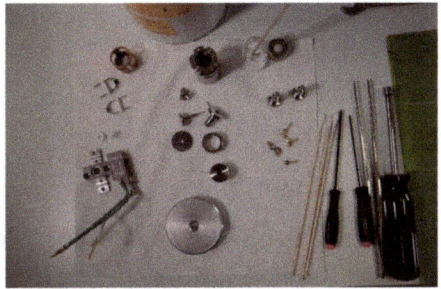

Figure 9.7 Images of Agilent ion source cleaning: in the left image taking the ion source apart; in the right image are the tools used, the ion source components that cannot be cleaned (on the left of this image), the components that are 100% metal and can be cleaned (in the middle) and the screws (on the right) can be cleaned if needed. Courtesy of Anthias Consulting Ltd.

- Once finished, remove any visible signs of the microgrit with a lint-free cloth (keep for next time).
- Place the parts in a clean beaker, cover with dichloromethane and sonicate for 5–10 minutes.
- Move the parts to another clean beaker, cover with acetone and sonicate for 5–10 minutes.
- Move the parts to another clean beaker, cover with methanol and sonicate for 5–10 minutes.
- Move the parts to another clean beaker and dry in the GC oven for 30–60 minutes at 100 °C, then cool the oven and components.
- Wearing clean lint-free gloves, put the ion source back together.
- Reinstall it and pump down the MS.
- Ensure the turbo is back up to 100% before leaving the instrument to equilibrate.

9.4.4.4.1 Filament Replacement

Ion source filaments are a consumable, they have a limited lifetime and will need replacing when:

- They break (no ionisation error).
- MS sensitivity reduces that cannot be explained by a dirty ion source. This can be diagnosed by switching to the other filament, if present and working.

If the MS has two filaments, the tune file can be switched to use the other filament, retuned and then used. Next time the MS is vented the filament should be replaced as described in the first part of ion source cleaning.

9.4.4.5   Mass Analyser

9.4.4.5.1   Quadrupole MS

Most quadrupole mass analysers do not require any physical cleaning or maintenance, but some MS instruments have pre-quads which do need physical cleaning. There is usually an option of thermal bake-out, which is raising the quadrupole to a temperature higher than that used in the method, to remove any contamination.

9.4.4.5.2   Ion Trap MS

Ion trap mass spectrometers usually requires physical cleaning to remove contamination, as the ionisation and ion trapping is all performed within the same area. There is usually an option of thermal bakeout that is raising the ion trap to higher temperatures to remove some contamination, but physical cleaning first is best.

9.4.4.5.3   Time-of-flight MS

TOFMS mass analysers rarely require any form of maintenance.

9.4.4.6   Ion Detector

The electron multiplier (EM) or microchannel plate (MCP) will need periodic replacement as it contains a finite number of electrons. Its lifetime depends on the applied voltage and therefore the number of electrons released, along with the number of ions/electrons striking it from the mass analyser (sample concentration). The maximum voltage may be 2000 or 3000 V, type and instrument dependent, and the voltage currently being used can be seen in the tune report.

It is worth noting, that once a high voltage is reached, for example 2800 V on an EM, there are few electrons left in the ion detector and therefore the maximum voltage will rapidly be reached. This is the time to order a new one, ready for replacement on an instrument that is frequently used.

Replacement of the ion detector is instrument dependent. On some MS instruments it is very easy, for example the Agilent GC-MS electron multiplier, after venting the MS, the old EM is unclipped and the single wire disconnected, the new EM is then removed from the packaging, wire attached and it is clipped into place (gloves must be worn!). Other instruments, usually those with microchannel plates may require an engineer to replace it.

9.4.5   PC and Software Maintenance

The data system and PC also require routine maintenance.

- The hard drive (HD) becomes too full of data files:
  o Regularly back-up data.
  o Regularly archive data.
- The HD is fragmented:
  o Clean out unwanted files.
  o Defragment the HD.
- The HD dies, in advance ensure you:
  o Back-up methods and data.
  o Ghost the computer set-up.

9.5   How Do I Know That My Maintenance Is Successful?

In Section 9.2.1 benchmarking was discussed to determine the performance of the instrument before maintenance. After maintenance this benchmarking should be repeated, and the results compared.

9.5.1   How Do I Ensure My Instrument Is Working Correctly After Performing Maintenance?

The steps should be:

1. Check the vacuum level, is it the same as before?
2. Perform an air and water check to see if there are any leaks before heating the GC.
3. Run a GC-MS system blank with no injection, to purge the system before tuning.
4. Once the oven is cool again, perform MS tuning and check the report.
5. Perform further, various blank runs:
   o GC-MS instrument blank with no injection to check for contamination of the system.
   o Solvent injection to check cleanliness of the solvent and autosampler.
   o Various other blanks to check other parts of the method, depending on the set-up.

6. Analyse a conditioning sample to mop-up any active sites – this could include matrix or be 2*x* the highest concentration standard. Conditioning samples are useful, but are not used in all industries.

7. Analyse the sensitivity test sample (described below), system suitability sample (discussed in Chapter 6) or the calibration standard (discussed in Chapter 8) to check the method and instrument chromatography and sensitivity. This may require a couple of injections to condition the system if a conditioning sample is not used.

8. Re-calibrate to check linearity.

9.5.2   Sensitivity and Performance Checks

Sensitivity can be checked with a target analyte from the application, or a more general compound can be used such as octafluoronapthalene (OFN).

OFN is a compound that is generally used for demonstrating GC-MS detection limits, and is also used by GC-MS manufacturers to report instrument limits of detection specifications.

It is an inert, non-polar, semi-volatile chemical that produces a sharp GC peak and a 70 eV EI mass spectrum with a dominant molecular ion (which represents 32.4% of its generated ions). It also has low relative abundance of isotopic isomers around the m/z 272 molecular ion. Additionally, the vacuum background noise in the m/z 50–300 mass range has the lowest value at m/z 272.

Typical OFN specifications for commercially available GC-MS instrumentation are a signal to noise ratio (SNR) greater than 100 using root-mean square (RMS) calculations for 1 pg OFN in full scan mode for $m/z = 272$, and/or a SNR greater than 100 (RMS) for 100 fg OFN in SIM mode for $m/z = 272$.

Other compounds for some applications have detection limits much less than this, but it is a useful compound to assess the performance of a GC-MS to assure the overall performance is maintained. Hexachlorobenzene (HCB) is also used by some manufacturers.

9.5.3   Maintenance Log

All laboratories should have a maintenance log book for each instrument. Ensure that all maintenance and post-maintenance checks are logged, along with any problems or physical changes made to the instrument. An example of a maintenance log can be seen in Figure 9.8 below.

Date	Check Cylinder pressure	Change cylinder	Replace gas traps & filters	Replace inlet septum	Replace inlet liner & o-ring	Replace liner seal (if suitable)	Clean inlet body	Trim column & replace ferrule	Review foreline pump	Electron Multiplier voltage	Repeller voltage	Ion focus voltage	Clean ion source	Replace filaments	Refill PFTBA calibration vial	Notes
29 Feb 2019	✓		✓	✓	✓		✓	✓	✓	1445	18					Maintenance done - IJ
15 Mar 2019	✓	✓														New Helium cylinder installed - IJ
30 Apr 2019													✓			Maintenance done - IJ
15 May 2019	✓			✓	✓	✓	✓	✓								Maintenance done - IJ

Figure 9.8   An example of a maintenance log for a GC-MS instrument.

This is very useful when deciding when maintenance needs to be next performed on a certain area and when troubleshooting!

9.6   Conclusions

- Use a log book for each instrument to record any maintenance and problems.
- For each instrument set up, determine all modules used (*e.g.* the inlet), all parts of that module that require maintenance and the frequency for each method.
- The method validation process should determine the maintenance frequency for each instrument part and the consumables for the application.
- Set up a maintenance schedule, not only for each instrument, but for each method if different methods are used on that instrument.
- Perform the maintenance as defined in the schedule.

The next chapter is troubleshooting. Many problems which are addressed in the troubleshooting chapter could be caused by a lack of maintenance.

10 How Do I Troubleshoot a Problem on My GC-MS?

10.1 Introduction

Many GC-MS problems can be prevented if the instrument is maintained routinely and looked after properly. However, the instrument may start to display problems that an analyst equipped with the right tools and understanding of the instrument, can identify, locate and correct with a minimal amount of effort. Troubleshooting is a skill that becomes easier with increased experience, knowledge and practice, if nothing ever went wrong, the authors of this book wouldn't have all of this experience to share!

There could be several parts of the system in which a problem can occur which will usually result in some type of chromatographic errors. It is possible for the same problem to be caused by several different parts of the system. It is useful in the first part of troubleshooting to identify the different modules in your system. Draw a diagram showing the carrier gas flow path and the sample flow path as a plan (see Figure 9.1).

Gas Chromatography-Mass Spectrometry: How Do I Get the Best Results?
By Diane Turner, Mathias Schäfer, Steven Lancaster, Imran Janmohamed, Anthony Gachanja and Jason Creasey
© Diane C. Turner, Mathias Schäfer, Steven Lancaster, Imran Janmohamed, Anthony Gachanja and Jason Creasey 2020
Published by the Royal Society of Chemistry, www.rsc.org

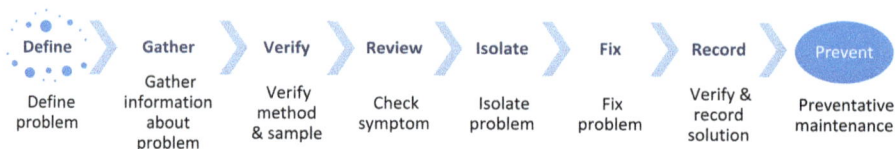

Figure 10.1 The troubleshooting process employed to identify a problem in the system. Courtesy of Anthias Consulting Ltd.

10.2 What Is the Troubleshooting Process?

A logical and controlled troubleshooting procedure will quickly and accurately identify the source of the problem. This will result in the fastest, easiest and most complete solution to the problem, as outlined in Figure 10.1.

10.2.1 Always Be Prepared to Deal with Problems

When the instrument is working well, it is important to collect and store information on its optimum settings. Analyse a reference standard of a known mixture (calibration standard, system suitability standard, *etc.*) and save a copy of the chromatogram for future review. Observe the noise trace – specifically have a good idea of where the baseline should be at a certain oven temperature. Keep a note of all conditions for the method used in both hard and softcopy. If possible, store the reference standard in the freezer to use again when needed. It is useful to have a folder next to the instrument with a print out of each of these, for quick and easy comparison.

10.2.2 Define the Problem

You need to know your system so that you can identify a problem when it is small!

- What does it sound like: autosampler/GC/MS?
- What does it smell like?
- What does it look like?
- What should your vacuum be?
- What should your tune results look like?
- What should your chromatogram look like for a particular analysis?

Always remember, if you don't know there is a problem you can't fix it and don't try to fix a problem that isn't there!

10.2.3   Gather Information About the Problem

The first step is to gather detailed information about the problem from previous operators, the instrument log book and sample log or sequence. Review the maintenance log to see when and what was the last maintenance performed. Other things to look at are:

- The previous samples and chromatograms that were run can provide information on the cleanliness of the system.
- Review tune and leak check reports.
- Check the instrument control panels and software for any errors reports or messages.

90% of problems occur due to the last thing that was done to the system!

10.2.4   Verify Your Method and Sample

It is important to quickly observe and eliminate any human errors arising from the use of the wrong method and sample.

- Confirm if the correct method has been loaded and used for the analysis, ensuring that none of the parameters in the method have been changed.
- Confirm that the correct sample has been analysed (look in the auto sampler tray and see if the sample is in the correct location), if there is enough sample in the vial, maybe a blank has been analysed instead of a standard.

10.2.5   Check the Symptoms

Review symptoms on the instrument and observe if the problem continues to be present and if it is getting better or worse.

Review and compare to other instruments in the lab to see if they experience similar problems as it could be associated with a common set up, for example a contaminated gas supply could show higher baselines in multiple instruments.

10.2.6   Isolate the Problem

The next step is to isolate the problem, identify and narrow down possible sources to one or more modules of the instrument or steps in the sample collection and analysis such as sample collection, sample preparation, gas supply, gas plumbing, autosampler, inlet, column, detector and data system. Attempt to remove the source of the problem and retest. For example:

- If it could be the autosampler, try performing a manual injection of the same sample.
- If it could be the inlet, try injecting into another inlet or perform a manual on-column injection.
- If there is a leak that can't be found, isolation of the MS by venting, removing the column, use of a blanking ferrule, pumping back down and retesting could help.

10.2.7   Fix the Problem

Once you have identified and isolated the problem to a particular area, fix the problem(s). However, only attempt to fix a problem that is recommended by the manufacturer. Any other problems could void any warranty or exacerbate the problem.

- Fix one potential problem at a time and retest after each to ensure it really was a cause. Problems could be caused by only one source or multiple sources, don't stop until you are sure you have fixed them all.
- If you have isolated the problem to multiple locations, try the quick, easy and cheapest solutions first. For example, if you have a choice between changing the septum and liner and cleaning the MS ion source, change the septum and liner first. If you cannot see any difference, then attempt to clean the latter.

If the problem is something that warrants a visit from a service engineer, prior to his visit discuss with the engineer what you think the problem could and could not be, plus what you have attempted so far. This can help to ensure that they bring the required parts for fixing the instrument.

At times, certain problems are very persistent and it can be very difficult to find a solution. Take a step back and re-evaluate what you have done so far. Discuss the problem with others who can help find a solution. Or even just discussing with someone else can help to clarify things for you and give you new ideas. Refer back to the plan of your instrument, to make sure that you haven't missed anything.

10.2.8 Verify and Record

Once you have fixed the problem, verify the result by running multiple samples, system suitability checks, sensitivity checks, blanks, and so forth to ensure that it is not a temporary solution.

If more expensive parts or consumables have been replaced that did not in the end relate to solving the problem, you can either re-install the original parts or save them to use again.

It is important that the all entries are recorded in the maintenance logbook – especially those that fixed the problem. Write it up properly – the symptoms, the causes and the fixes, as it might be useful years down the line, or a colleague could have the same problem next week!

10.2.9 Prevent

If the problems are related to issues that regularly appear on the instrument, it may be important to create a task, for example preventative maintenance to reduce the likelihood of re-occurrence.

Discuss with other users of the instrument to ensure that the problem doesn't re-occur and that the knowledge and experience gained is shared.

10.3 Useful Tools for Troubleshooting

It is useful to have your manufacturer's instrument manual along with the following diagnostic tools at hand:

- New syringes, septa, ferrules, inlet liners, and other consumables
- No-hole ferrules for the MS transfer line and the GC inlet
- 1/8" blanking nuts
- On-column syringe needle
- Reference/system suitability/sensitivity check standard
- Flow meter
- Electronic leak detector
- Reference column
- Vial of methanol – useful for checking flow through the column and cleaning.

10.4 A Breakdown of Problem Areas

The instrument and method can be broken down into the individual modules. Some symptoms can be related to only one module whereas others could come from multiple modules. It is useful to know what problems can occur in each step of the sample analysis process. Again, when troubleshooting it is a good idea to draw a diagram of all the inputs and outputs of a method into these modules or areas.

10.4.1   Problem Areas: Samples

There are various problems that can occur with the samples. It is important that there is an understanding of the sample, analytes and matrix and their chemistries.

- Samples can be unstable – discrimination, breakdown or reactions can occur. Consider preservation, storage, derivatisation and shaking before injection.
- Samples can evaporate from the vial on the autosampler – consider the types of vials, septa and vial storage in the tray (cooling tray).
- Foaming samples or emulsions being formed, consider using anti-foaming agents particularly for purge and trap.
- If there are variable amounts of salt in the sample or changes in pH, consider normalising them.
- If the samples are too high in concentration they could contaminate the instrument – run blanks after these and consider a pre-screening method, for example on a GC-FID (flame ionisation detector), especially if purge and trap is used, try with headspace first.
- If the samples are too low in concentration consider concentrating the extract either with manual blow-down or with a large volume injection. Don't inject more without checking the vapour volume and that it fits in the GC inlet liner!
- If the samples contain particles which could block the syringe consider filtering, but check that analytes are not lost nor contaminant peaks added. Alternatively, change the position of the syringe tip when collecting sample from the vial.
- If the sample is too low in the vial – consider using vial inserts or a tapered vial.

If a problem has been identified relating to the:

- Sample vials:
  - Curved corners are better as they are stronger and less likely to 'blow out' if heated, they are also easier to place back into the autosampler tray.
  - Watch for variations in shoulder and neck height.

- Vial caps:
  - o For crimp caps ensure the cap is level and not at an angle, then ensure the cap does not spin once crimped otherwise it will leak.
  - o For screw caps don't over-tighten and damage threads, these are very useful if the sample needs to be kept after analysis and so should be re-capped then stored.
- Vial septa:
  - o These are usually a silicone type with PTFE base and should be selected to be resistant to the sample.
  - o For crimp caps use thicker septa and screw caps thinner septa.
  - o Ensure they are not stored near solvents as they will absorb them and try not to store them in plastic bags as they are porous and the bags contain plasticisers!
- In general:
  - o Ensure all vials, caps and septa are rated to the temperature used.
  - o Condition and store them in glass jars for low limits of detection (LODs).

10.4.2   Problem Areas: Gases and Plumbing

The common problems associated with gases are mainly leaks and the presence of impurities, such as:

- Oxygen can damage column phases and reduce the performance of detectors such as electron capture detectors (ECDs) and MS.
- Water can damage some column phases and reduce the performance of MS.
- Hydrocarbon impurities can cause background contamination especially in FID and MS.

Symptoms of leaks and impurities are:

- Loss of carrier gas.
- Higher baselines reducing the signal to noise ratio (SNR) of the target analytes.
- Damage to the GC column causing activity and peak tailing.
- Make the MS vacuum worse, reducing the sensitivity.

To reduce the likelihood of these problems:

- GC-MS should be leak-checked daily and always before use or tuning.
- Fittings should be leak-checked regularly and always after changing or altering anything.
- Gas traps and filters should be installed and maintained.

10.4.3   Problem Areas: Autosampler

An autosampler can be a common place for errors to occur, the exact problems are dependent on the type of autosampler, what it does in terms of automated sample preparation and injection and if, for example, the carrier gas passes through it. For a list of common problems in each type of autosampler, please refer to Chapter 9. Possible autosampler issues are:

- Leaks – if the carrier gas passes through it, or it uses any gases.
- Contamination and carryover from high-concentration samples and/or methods not properly optimised.
  - Always consider a less sensitive screening technique for very sensitive autosamplers, such as purge and trap.

10.4.4   Problem Areas: Inlets

As discussed in previous chapters, inlets perform three jobs, as they are where the (dirty) sample is introduced and transferred to the column, resulting in many problems being caused by a lack of inlet maintenance or the sample introduction method not being properly optimised. Problems can also be caused by, for example, a change in the sample solvent, without considering the effects on the method. For example changing from hexane to methanol the vapour volume will be much larger, plus if a splitless injection is used, the initial oven temperature may now be too high for the solvent effect to occur, resulting in poor early peak shapes, a drop in the peak size and contamination of the inlet producing carryover.

Vapour volume: does the volume injected of a particular solvent fit into the inlet liner with the method parameters selected (Chapter 2)? Exceeding the liner capacity (also known as flashback) will result in:

- Sample being lost *via* the split exit and/or septum purge line thereby reducing the peak size.

- Ghost peaks from the analyte molecules being pushed outside of the liner and being transferred to the column later. Sources of ghost peaks include the septum, carrier gas in line and the dead volume around the outside of the liner and down the split line in a splitless injection.
- Carry over from sample going back up the carrier gas line then being analysed in subsequent runs.
- Contamination of the liner on the outside, the inlet body, the liner seal, the carrier gas, septum purge and split lines, the head of the analytical column. Maintenance will be required to fix each of these.
- Overall it will result in a loss of sensitivity, poor peak shape and poor reproducibility. Always, always check the vapour volume using a free calculator when developing the method and before changing the solvent type, injection volume or liner volume.

Inlet temperature: this should be optimised for the analytes in the sample.

- Too cold leads to mass discrimination of the higher molecular weight (MW) analytes.
- Too hot leads to thermal degradation of thermally labile analytes (particularly in splitless injections), and high MW dirt being transferred into the column.
- Inlet heaters can also break resulting in no heating of the inlet.

Amount of sample transferred onto the column:

- Split injection: split ratio will affect the proportion of sample transferred to the column.
- Splitless injection: this will be affected by the splitless (purge) time. Opening the split exit too early results in loss of sample. Opening the split exit late will result in tailing peaks, particularly of the solvent peak.

Analyte focusing on the column: see below under columns.

Activity: inlets are the main source of activity in a GC. Activity mostly affects active (polar) compounds resulting in poor peak shapes, for example tailing peaks; is a site of catalytic breakdown and irreversible adsorption causing a drop in sensitivity for an analyte or even complete loss for low-concentration active analytes. Activity can come from numerous sources within the inlet:

- Liner used: deactivated liners must always be used for active compounds, especially when splitless injection is used as the molecules spend longer in the liner with more opportunity for interaction.
- Dirty liner: dirt causes activity, so even if a deactivated liner is installed multiple injections, especially of dirty samples, will gradually increase activity within the inlet – perform maintenance!
- Dirty inlet or inlet seal: clean or replace.
- Poor column trimming: increases the surface area of the end of the column, which contains active silanol groups.

Column positioning: if the column is installed too far into the inlet it can lead to:

- loss of sensitivity;
- poor peak shapes – broadened or split;
- discrimination.

Septum and septum purge flow: the correct septum for the application must be used and installed correctly:

- *Temperature rating*: otherwise excess bleed and degradation will be seen.
- *Resistance to solvent*: otherwise the solvent on the outside of the syringe needle will damage and dissolve the septum, causing excess bleed and leaks.
- *Contamination and bleed*:
  o Always use gloves when installing.
  o Before use store in a solvent-free location in appropriate packaging.
  o With many injections, sample on the outside of the syringe needle will accumulate in the septum, ultimately leading to ghost peaks and carryover.
  o Use septum purge to purge septum bleed and contamination to waste. Default is 3 mL min^{-1} but some instruments allow this to be changed or programmed.
- *Installation too loose*: it will leak!
  o Producing an additional split flow upon injection, resulting in reduced sensitivity of the analytes.
  o The inlet pressure required for the column flow may not be reached, resulting in changes in retention times or the inlet will switch off.

- *Installation too tight*:
  - o Can also result in leaks as not all surfaces will be in full contact with the inlet head.
  - o Can produce excess coring of the septum when the needle is pushed through.
  - o Can even bend the syringe needle!
- *Septum coring*: small particles of the septum are broken away by the syringe needle, reducing the thickness of the septum at that point, leading to a shorter life. The particles will drop into the inlet liner and be analysed producing individual peaks in the chromatogram. In a liner with no packing they can drop onto the inlet seal and block the split flow. Coring can be minimised by:
  - o Using the correct syringe needle: cone-tipped is best, square-tipped (*e.g.* solid phase microextraction (SPME) fibre needles) are the worst and requires pre-piercing of the septum, tapered are not so good as they cut rather than push through the septum.
  - o Using soft septa: high temperature septa are harder, therefore only use if required by the inlet temperature.
  - o Using a long-life septum such as a Merlin Microseal™, or an inlet septum-less head. For both ensure the correct syringe needle (23 gauge and cone-tipped) is used to avoid damage.

Inlet flows: there are many types of flows through the inlet:

- *Total flow* into the inlet: this is the sum of the split flow plus column flow plus septum purge flow. If the total flow is higher than the sum of these three flows then there is a leak! This is a good, quick check.
- *Split flow*: as well as affecting the amount transferred onto the column, as discussed above, this will affect peak shapes, in particular causing a tailing solvent peak if there is a split flow problem. The split flow can be measured with a flow meter at the split line exhaust. If it is not reaching the required flow there will probably be:
  - o A blockage on the liner seal, usually caused by septum coring (see above).
  - o Blocked split line, see Chapter 9.
  - o Blocked split line trap – replace as discussed in Chapter 9.
- *Liner flow*: equal to the column flow in a splitless injection and the column flow + split flow in a split injection. Flow through the liner can be restricted if any liner packing is compressed, for example if you pack it with wool. This wool affects the transfer of analyte molecules to the column, resulting in reduced sensitivity.
- *Purge flow*: as discussed above under septum.

10.4.5   Problem Areas: Columns and Column Oven

Most column problems are related to peak shape, contamination and leaks, for example:

- Fronting, tailing, broad or split peaks caused by overloading of the column, poor transfer to the column, activity in the column, poor column installation, dead volumes in column connections, flow through the column is incorrect or not maintained.
- High baselines caused by damaged columns, leaks, contamination (coming from before the column) which is volatile and does not condense on the head of the column, contaminated detector.
- Additional peaks caused by carryover from previous runs and contamination from modules before the column, condensing at the head of the column to produce peaks.

The main areas and concerns with the column are:

- *Focusing on the head of the column*: this is very important in split-less injections, in which the transfer of the analyte molecules to the column is slow. Analyte focusing is produced using two different techniques:
  - Cold trapping: the stationary phase must be thick enough and the initial oven temperature low enough for the analyte molecules to condense, be retained and focused.
  - Solvent effect: the solvent is condensed on the head of the column and the volatile analytes dissolve in it.
    - The initial oven temperature should be 10–20 °C below the solvent boiling point for this to occur.
    - The stationary phase polarity should match the solvent polarity, otherwise the peaks will split. If not matched, use a pre-column of the correct polarity for the solvent.
- *Separation and resolution*: the chromatographic resolution of peaks should be optimised in method development then monitored through system suitability. Try to select the two peaks with the least resolution (a critical pair) and monitor these. A reduction in resolution can be from:
  - Reduced amount of stationary phase in the column: columns bleed all of the time and especially with every temperature cycle. Over weeks/months/years, even without column trimming, the amount of phase will reduce, ultimately resulting in a loss of resolution indicating that the column needs to be replaced.

o Change in carrier gas velocity: the method should have been developed to use the optimal velocity for the separation. A change in the velocity is usually seen by a change in the retention time. This could be due to the amount of stationary phase changing through column trimming or bleed (see above) or a column blockage or leak (see below).

- *Final column temperature*: the final temperature and hold time of the GC oven should be such that:
  o It doesn't exceed the column maximum ramping temperature (see below).
  o Analytes elute on the temperature ramp to minimise band broadening and keep the peaks as narrow as possible.
  o It is high enough and/or held long enough that all molecules transferred to the column are eluted in that run (or during post-run or backflushing if these are used). Failure to do this will result in very broad peaks, owing to excess longitudinal diffusion, eluting in the following run.
  o Control what is transferred onto the column through the inlet temperature.
- *Column bleed*: excess column bleed is mainly caused by a lack of care with the column and the system.
  o When not in use store the column in the box to avoid damage and seal the ends with a septum to reduce contamination.
  o When installing, condition according to the guidelines provided in Chapter 9, to remove damaging oxygen and water and to reduce bleed. In the GC oven, hang on the column hanger correctly, ensure the column doesn't become trapped, tightly bent or touch the oven walls.
  o Avoid transferring dirt, high molecular weight matrix, inorganic acids and bases to the column from the inlet – sample preparation and inlet temperature are important.
  o Trim the head of the column by 1–2 inches if poor peak shapes or excess bleed is seen.
  o Leaks: see below.
  o Use within the specified:
    ■ Maximum isothermal temperature for the inlet, transfer line and detector, when conditioning or using an isothermal temperature program.
    ■ Maximum ramping temperature within the GC oven, holding for a maximum of 20 min at this temperature.
- *Excess column bleed*: high levels of column bleed will:
  o Increase the baseline resulting in a reduction in the SNR for analyte peaks.

  o Dirty any GC detectors and the MS ion source, it may even get through to the mass analyser and detector, increasing their maintenance.

  o Result in areas of column with no phase, producing active sites to cause tailing peaks and potentially a loss of analyte molecules, reduced sensitivity and resolution.

- *Leaks*: a leak can affect the health of the column as well as affect the chromatogram.

  o Check for leaks in the system, to prevent oxygen damage to the stationary phase, especially when at high temperatures, which at a minimum will result in higher baselines and a lower SNR.

  o Leaks can occur at any time, therefore, keep GC oven temperatures low (<60 °C) with the gas flow always turned on at 1–0.7 mL min^{-1} (absolute minimum of 0.5 mL min^{-1}) when not in use.

  o A leak at the head of the analytical column (in the inlet or connection with a pre-column) will reduce the flow through the column changing the retention times. It can also reduce the amount of sample transferred to the column (see inlets above), affecting peak size.

  o Seeing no peaks in the chromatogram could be the result of a broken column. Visual inspection can often be used to spot this. If not, check the flow through the column by disconnecting from the detector and placing the end into a vial of methanol to look for bubbles. It is useful to do this upon column installation before connecting to the detector.

  o A crack in the column, causing a leak, is the most difficult problem to spot and is usually suspected when every other potential cause has been eliminated. Prevent the possibility by careful column care and not damaging the polyimide coating.

10.4.6   Problem Areas: GC Detectors

There are many problems that relate to specific GC detectors, however the more common problems are:

1. *Quenching (flames out)*:
  o Incorrect detector parameters such as the flows or temperatures used – optimise.
  o Dirty detector – clean.
  o Leaks – fix.
  o Solvent volume – evaluate or optimise parameters for volume injected.

2. *Spiking, noise or drift*:
   o Contamination and dirty detector parts – clean.
   o Check gas filters and traps are used and maintained for detector gases.
   o Column positioning: if the column is installed too far in it interferes with the detector operation and may damage the end of the column.
   o The sample.
   o Electrical problem: check signal connections and insulators.
3. *Poor sensitivity*:
   o Column positioning: if it is too far out there is a long distance for analytes to travel into the detector resulting in reduced sensitivity, particularly for less volatile analytes.
   o Acquisition rate: too high reduces the S/N ratio. Most GC detectors only require 5–10 Hz, unless GCxGC is used.
   o Operating parameters are not optimised.

10.4.7 Problem Areas: Vacuums

Vacuum problems can best be diagnosed by observing changes to the normal vacuum readings of your system and listening regularly to the noise of each pump. Turning on the vacuum pumps after the system has been vented or after a power cut is the most likely time for a problem to occur. Most problems are caused by leaks, see Section 10.5.1.

Vacuum problems occurring when pumping back down after venting:

- Mechanical pump not switching on: check it is connected to a power source and switched on.
- Mechanical pump gurgling for more than 1 min: there is a large leak. Switch back off and check all connections.
- High vacuum pump not switching on:
  o Not enough vacuum produced by the mechanical pump for the high vacuum pump to turn on – it requires 10^{-1} to 10^{-2} Torr, check mechanical pump is ok.
  o Not connected to a power source.
  o Fan not working.
  o Pump failure.
- High vacuum pump not reaching 100%:
  o How long does it usually take? Soft start takes much longer and it will hold at a certain revolutions per minute before increasing again.

o A largish leak: vent, check all connections then try again.
- System seems to be working well, but air/water check indicates a problem:
  o Nitrogen should always disappear faster than water, as water sticks to metal surfaces and takes longer to purge.
  o After pumping down, even if both air and water are still high:
    ■ Water greater than nitrogen = no leak, but will take time to purge both.
    ■ Nitrogen greater than water = a leak, identify and fix before leaving to purge.

Vacuum system not pumping efficiently:

- Sensitivity is reduced.
- Can be checked by turning on the calibration gas in manual tune then turning off, the mass spectrum of gas should disappear quickly as it is pumped out!
- Caused by:
  o leaks;
  o too high column flow;
  o leak in the vacuum system post-ion source, for example hoses between the high vacuum pump and mechanical pump.

Mechanical pump problems:

- Gurgling: a large leak! Switch off the vacuum and check all connections in the instrument, starting from the pump and working backwards. Note: when the pump is first switched on there will be gurgling for the first 30 seconds. If it continues for more than a minute, there is a problem!
- Smallish oil leak: determine the source of the leak. Vent the MS in a controlled way. Check all fittings are closed. If it is over-filled, remove some oil as if the oil was being changed, see Chapter 9. Replace any o-ring seals as needed. Service the inside of the pump if required.
- Pump has dumped all of its oil: immediately turn off the MS (do not vent). Service the pump.
- Always clean oil from the pump and the pump tray so that it is easy to spot further leaks.

Turbo pump problems:

- Metallic 'ting' sounds are heard: bearings could be seizing causing spinning blades start to hit fixed blades. Best to replace bearings or source a re-conditioned pump.
- Turbo is over-heating: check pump cooling fan and replace if necessary. If fan is working well, this is another sign of the bearings starting to seize.
- Self-destruction of turbo (very loud noise): bearings have seized, pump must be replaced.

Diffusion pump problems:

- Oil vapour seen in mass spectrum:
  o MS not vented correctly – review the procedure.
  o Leak around the pump – fix.
  o Perform maintenance of the diffusion pump oil.
- Vent MS, completely switch off and remove power supply cable. Manually clean MS surfaces very carefully wearing gloves with a lint free cloth and a suitable solvent. Only wipe visible oil and do not clean delicate parts (*e.g.* mass analyser), only larger metal surfaces. Do not spray MS with solvent, put directly onto the cloth. Do a final clean with methanol. The background will take a long time to reduce!

10.4.8 Problem Areas: Mass Spectrometer

Most mass spectrometer problems are observed from the air/water and tune reports and on examination of the chromatograms and mass spectra. These include:

- *Air and water leaks*: see Section 10.5.1.2.
- *Contamination*: from pump oil vapour, diffusion pump oil, calibration gas ions, GC – see Section 10.5.2.
- *Low sensitivity*: usually maintenance is required:
  o Dirty: clean the ion source.
  o Replace the filament.
  o Check the detector (electron multiplier (EM) or microchannel plate (MCP)) voltage – see Chapter 9.
  o Check the vacuum.
  o Check for leaks.
  o Ensure the GC column flow is optimal and not too high.

- *Poor mass resolution or mass accuracy*: check the instrument is tuned, clean and that there is plenty of calibration gas.
- *Shorting within the ion source*: indicated by strange tune results or it won't tune. Check and clean the ion source, if just cleaned – check the wires are the correct way around and that the ion source is installed correctly.
- *Poor tuning or it won't tune*:
  o No calibration gas: it has run out or the gas valve is stuck closed.
  o Shorting: check ion source is installed correctly – see above.
  o Contamination: see above.
  o Low sensitivity: see above.
- *MS not heating*, for example the ion source or mass analyser:
  o Ensure it is pumped down and has a good vacuum.
  o Check applied temperatures.
  o Heaters may need replacing.

10.4.9   Problem Areas: Computers and Data System

Personal computers and data systems can also be a source of potential problems. A full hard drive can stop the collection of chromatographic data. However, with modern day hard drives with terabytes of data, this is now becoming less critical, but routine disk clean-up and defragmentation is important. Hard drive failures commonly occur and hence it is important to ensure methods and data are stored and backed up.

10.5   Common Problems

Some of the most common problems that can occur on the GC-MS are as follows:

10.5.1   Leaks

There are two types of leaks: leaks out of the system and leaks into the system. Each of these may only occur on one area of the system, but both might occur in other areas (not at the same time). It is important to follow carrier gas and sample flow paths and understand the vacuum system to quickly identify the causes of and fix leaks. The more fittings there are, the greater the potential for leaks, therefore minimise the number of connections as far as possible.

10.5.1.1   Leaks Out of the System

Leaks *out of* the instrument are losses of the carrier or detector gases that can occur anywhere between the cylinder and the detector, depending on the fittings and pressures used.

Leaks out of a system can be identified from:

- Retention time shifts, as there is a reduction in flow.
- Area count changes and a loss in sensitivity, as the leak will act as an additional split.
- The cylinder volume will decrease faster than usual – always keep a weekly record of the volume, as discussed in Chapter 9.

The exact location of a leak should first be identified with a leak detector (set-up for the gas type). Liquids, such as soap solutions, should be avoided as they will cause contamination and corrosion, many contain organic materials that can be detected by GC. Liquid leak detectors, such as snoop, can be used to look for leaks away from the GC, for example between the cylinder and the external fitting of the pressure control module (EPC) but should never be used on/in the GC itself.

Some of the typical sites are high-pressure systems and/or where there is a reduction in the tubing (including column) internal diameter (i.d.) Places to check are:

- Carrier gas supply: regulator, gas line fittings, tubing up to and including the EPC.
- EPC connections and o-rings, tubing to the inlet, split and septum purge lines, also check weld attachments of the tubing, *e.g.* to the top of the inlet, as these can crack.
- Inlet: septum, liner seal, base seal.
- Column: ferrules, inlet connection, press-fit and other column connectors, GC detectors.
- These don't occur around the MS as there is a vacuum!

NEVER start by tightening all fittings, as most will be fine and further tightening could now produce a leak. Always identify which fitting is causing the problem with the electronic leak detector and fix that, then repeat for further leak sources.

10.5.1.2 Leaks into the System

Leaks into a system are typically seen with a vacuum detector, such as GC-MS, or sensitive detectors such as an electron capture detector (ECD).

The typical locations for a leak into the system are:

- Column connections to the inlet, any press-fits and most importantly the transfer line.

- MS hardware parts, such as the vacuum system, for example, turbo, hoses, vent valve o-ring, calibration gas vial o-ring, door seal/o-ring, and so forth.

Leaks into a system can be identified from:

- High levels of nitrogen and water from air/water checks (possible to see in tune reports too).
- Higher baselines caused either by interference with the detector, or higher column bleed owing to oxidation of the stationary phase.
- Reduced sensitivity.

With GC-MS, the detection of a leak can be performed using a spray or vapour which contains ions whose mass is known. This could be:

- Argon gas at *m/z* 40
- Methanol headspace at *m/z* 31
- An air-duster spray containing a halogenated compound, for example:
  - *trans*-1,3,3,3-tetrafluoropropene (HFC-1234ze) *m/z* 114
  - 1,1,1,2-tetrafluoroethane (R134A) *m/z* 83.

The process to check for leaks is to:

1. Start the MS scanning from *m/z* 15 to 100/120 (dependent on gas used).
2. Spray the gas onto a potential leak site, see Figure 10.2.

Figure 10.2 Checking for a leak at the mass spectrometer transfer line using an air duster.

3. Watch the ions – if there is a leak nitrogen m/z 28 will go down and the gas ion will go up in abundance.
4. Slightly tighten the fitting and re-test, do not overtighten! Use the one-finger test: use only one finger on the wrench, if the nut moves continue, if not do not tighten further. If there is still a leak at that site, then that ferrule will need replacing.
5. If something needs replacing (*e.g.* septum), stop the MS acquisition, replace and then retest for leaks.
6. If the leak is fixed at that site move to the next potential leak site – remember that there may be multiple leak sources on the instrument.
7. Start from the MS (reaches ion source faster) and work towards the inlet.
8. At each site wait for the ions to reach the MS, on the GC inlet it could take up to 2 mins for gas to go through the column (check the column dead time on a flow/pressure calculator).

10.5.2   Contamination

Contamination is one of the most challenging troubleshooting problems because it can come from multiple sources, therefore it is best to work through the GC-MS system. Contamination can present in two different ways:

- *Presence of additional peaks in the chromatogram*: these have been focused on the head of the GC analytical column and separated as a chromatographic peak. Therefore, the source is from before the head of the analytical column, for example a sample, carrier gas supply, autosampler, inlet, pre-column or connection.
- *An increase in the baseline or noise in the mass spectra from a non-coeluting peak*: this contamination has not been focused on the head of the analytical column. Therefore, it is either too volatile or the source is from the column to the detector, for example volatiles in the carrier gas, column bleed, detector contamination, a leak or electrical noise (baseline only).

You should examine the chromatogram and mass spectrum to identify the type of contamination and therefore possible sources. Examine the ions in the peak mass spectrum and run a library search to identify the source. For contamination visible in the baseline also check the ions and refer to Table 10.1 for the potential source.

Table 10.1 Table containing the masses, type and potential source.

Masses (m/z) observed[a]	Compound classification	Potential compound source
18, 28, 32, 40, 44	Water, air, argon, carbon dioxide	Residual air and water air leak
31	Methanol	Cleaning solvents
91, 92	Toluene, xylene	Cleaning solvents
105, 106	Xylene	Cleaning solvents
43, 58	Acetone	Cleaning solvents
151, 153	Trichloroethane	Cleaning solvents
69	Foreline pump oil or PFTBA	Pump oil vapour or calibration valve leak
73, 147, **207**, 221, **281**, 295, 355, 429	Dimethylpolysiloxane	Septum or stationary phase
77, 94, 115, 141, 168, 170, 262, 354, 446	Diffusion pump fluid	Diffusion pump fluid
Masses 14 m/z apart	Hydrocarbons	Fingerprints, foreline pump oil
149	Phthalates	Plasticizers in tubing, vials, caps, samples

[a]Bold = key column bleed ions to look for.

Possible sources of contamination through the GC-MS system can be:

- Sample: sample prep, vial/cap/septum, or the sample itself.
- Carrier gas: impurities, leaks.
- Autosampler: syringe, wash solvents, needle guides, and so forth.
- Inlet: septum, liner, body, liner seal, split/purge/carrier gas lines, fingerprints.
- Column: bleed, fingerprints.
- Detector: dirty, cleaning solvents, fingerprints, electrical noise, oil.

Analysing different types of blanks can help to determine the source:

1. GC-MS instrument blank with no injection: checks inlet to detector plus carrier gas.
2. Manual solvent injection with a new syringe *versus* autosampler injection of the solvent: can help to identify or eliminate the autosampler. If seen in both it could be the solvent or the solvent has dislodged the contamination in the inlet owing to the vapour.

3. If the autosampler is something like a thermal desorber, heating just the cold trap *versus* analysing a clean TD tube can determine if the problem is pre- or post-cold trap.
4. Moving the column to a different inlet if installed or, performing a manual on-column injection of solvent can help to identify or eliminate the inlet.

Depending on the likely location of the problem:

- Analyse the solvent (isolate sample related problem) as described earlier.
- Replace consumable items (inlet, autosampler).
- Clean the inlet/autosampler if necessary.
- Trim the front end of the column.
- Physically clean the detector (ion source).
- Thermally clean the detector (ion source/mass analyser) and at the same time bake-out the instrument for 1 h (don't exceed column maximum isothermal temperature).
- Run another blank – has contamination been removed or improved?
- Re-condition the column overnight if necessary and thermally clean the detector for longer.
- Remember, cleaning will introduce more contaminants at first so a couple of blanks may need to be run!

Baseline increase could be:

- Carrier gas impurities:
  o Check other instruments using the same supply.
  o Check maintenance of traps (last time replaced?).
- Column bleed: bleed ions present?
  o Trim column – improved?
  o Recondition – remove from MS if possible.
  o Check transfer line temperature – above column maximum isothermal temperature?

Ion source dirty and needs cleaning, check tune report:

- large number of peaks in mass spectrum (Agilent >200);
- high repeller/push plate voltage (Agilent max. = 34.91 V);
- low abundance for the tune type;
- tune masses do not have smooth shapes (especially smallest 502 ion).

MS recently cleaned? Residual solvents?

- Ensure no leak is present.
- Purge for longer – overnight/weekend.
- Increase the column flow (do not exceed vacuum capabilities).
- Increase the oven temperature (do not exceed the column maximum isothermal temperature).
- Transfer line temperature on and correct?
- Ion source and mass analyser temperature on & correct?

Mass analyser dirty:

- lots of fragment ions in the mass spectrum;
- no leak;
- not column bleed ions;
- cleaning the ion source hasn't helped;
- not cleaning solvents.

Instrument dependent:

- Physically clean the mass analyser.
- Bake out the mass analyser.

Caused by:

- samples too high in concentration;
- lots of matrix;
- solvent delay being used.

Mechanical pump:

- Anti-suck-back valve not working.
- Seals degrading.

Diffusion pump:

- Oil level not correct.
- Oil dirty.
- System not vented in the correct manner.
- Mechanical pump not working properly.
- Leak in vacuum system.

Vent and inspect inside the MS:

- Oil droplets may be visible.
- Clean the instrument parts, where possible and allowed, with a lint-free cloth and give a final clean with methanol.

It is important to ensure that long-term solutions to contamination are addressed.

- Use high purity carrier gases and use and maintain traps and filters.
- Use of columns:
  o Condition not attached to the MS to prevent bleed contamination.
  o Always install the same way around so any involatiles stuck at the head of the column aren't backflushed into the MS.
  o Ensure the transfer line does not exceed the column maximum isothermal temperature.
  o Use low-bleed, MS rated phases.
  o Use within the temperature limits, remembering the maximum isothermal temperature for the transfer line.
- Maintenance:
  o Always wear gloves.
  o Gently wipe towards the end of the column with a cotton bud and methanol to remove finger oils.
- Samples:
  o Reduce the concentration getting to the detector.
  o Remove the matrix.

10.5.3 Sensitivity

To be sure that the GC-MS works well and is fit for purpose, good practice would be to run a method-specific system suitability test to check that there is adequate sensitivity.

As with contamination, a lack of or a drop in sensitivity can come from multiple places throughout the GC-MS instrument. A lack of sensitivity can be caused by:

- A leak into or out of the system – analyte is lost due to an additional split, baseline increased or detector de-sensitised.
- Contamination – as the baseline increases, reducing the signal to noise ratio.

- Lack of maintenance – dirty autosampler, inlet, column, detector.
- Discrimination losing volatiles or involatiles in various ways throughout the system.
- Thermal decomposition.
- Activity within the system – adsorption.
- Poor quality carrier gases.
- Column not installed or trimmed correctly.
- MS not tuned.
- Sample/standard: not prepared correctly, wrong concentration, evaporated.
- Injection: the full required volume was not injected into the system.
- GC detector or MS (ion detector) not set to the correct sensitivity.
- Band broadening.

10.5.4   Activity

Activity within the GC-MS system can result in the loss of particular compounds, usually more polar analytes. The sources include:

- sample preparation;
- inlet;
- column – analytical column or pre-column.

10.5.5   Retention Time Shifts

Peaks can all move earlier, all move later or individual peaks can shift, depending on the sample.
   Shifting earlier can be caused by:

- column aging or less stationary phase from trimming;
- temperature or pressure control.

Shifting later can be caused by:

- Leaks resulting in the flow being reduced.
- Partial blockages: flow not maintained.
- Temperature or pressure control.
- A new column is installed containing more stationary phase than the old column.

Changes in the retention time of individual peaks is usually the result of dirty samples with high levels of matrix. Matrix can sit on the column and depending on the composition of the matrix and the chemistry of the analyte, peaks can:

- Move earlier, as the matrix blocks the interaction sites, reducing the amount of reachable stationary phase so they are retained by the column for a shorter amount of time.
- Move later, as the matrix has a greater interaction with the analyte molecules compared to the stationary phase, causing them to be retained more and to move slower than usual through the analytical column.

10.6 Summary

When faced with an instrument or sample analysis problem don't panic! Be methodical and logical, don't just jump in – you could cause more problems!

- Check the logbook, study the chromatogram and mass spectra, check the tune, air and water reports.
- Decide on a plan.
- Simplify the instrument, reduce the possible sources of problems.
- Identify and repair, then test findings.

Above all, define and carry out regular maintenance in order to reduce the likelihood of having to perform troubleshooting.

10.7 Problems

Below are a few troubleshooting steps that are useful to consider in the troubleshooting process. For each problem in Charts 10.1 to 10.6:

1. Consider each part of the GC-MS instrument.
2. Compare the problem chromatogram against the reference chromatogram.
3. Define what is the problem?
4. Define what are the possible causes?
5. Define what are the possible solutions?

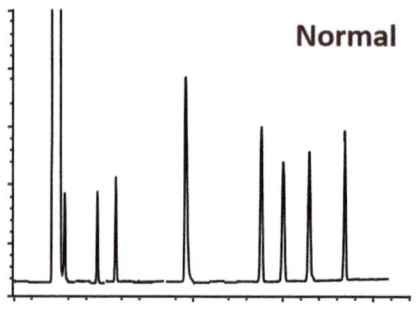 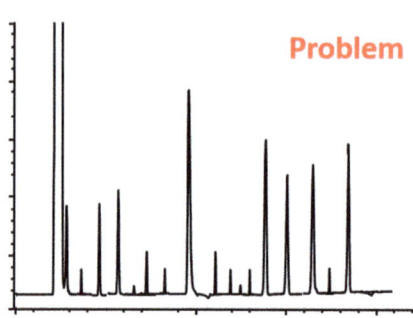

Chart 10.1

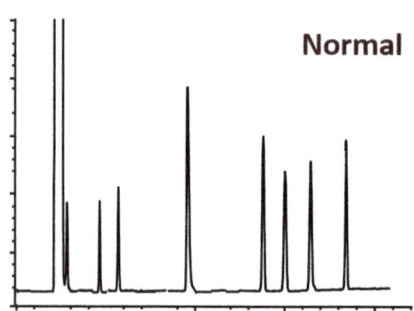 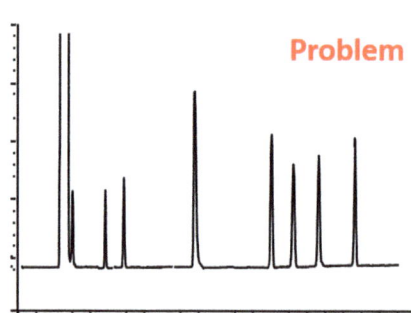

Chart 10.2

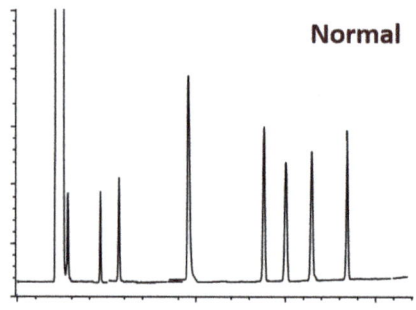 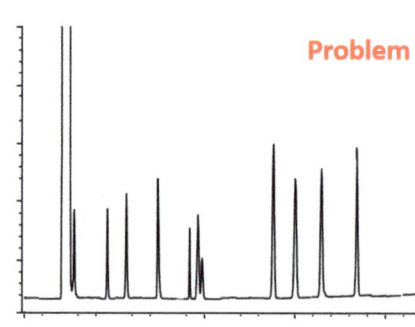

Chart 10.3

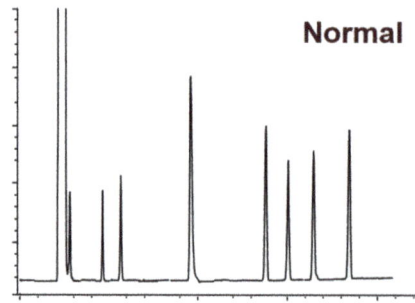 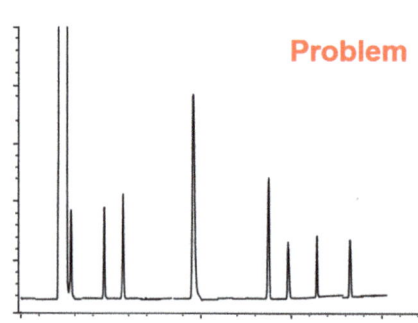

Chart 10.4

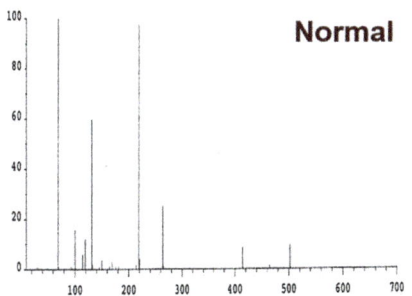 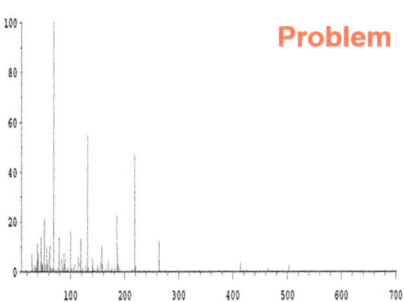

Chart 10.5

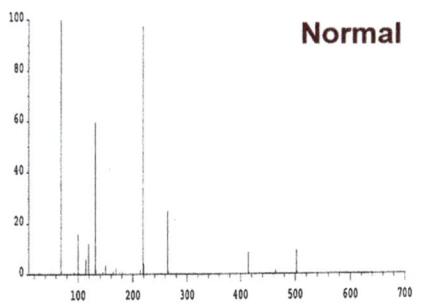 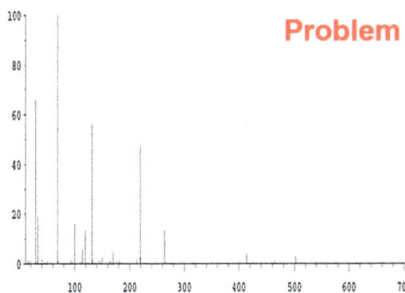

Chart 10.6

11 Conclusions

It is very rare for an analyst to be involved in the specification, acquisition, installation and use of an instrument. Therefore, the GC-MS in the lab may be an unknown quantity. If fortunate, a previous user of the instrument, someone involved in any of these steps, may be available to help and give lots of information. However, more commonly than it should be, an analyst may encounter an instrument and be left to get it working relatively alone. Whatever your experience, when first approaching your GC-MS instrument, get to know it.

- What type of gases does it use, what is their source and their purity?
- Trace the gas lines towards the instrument – what fittings are in the gas plumbing? Is it used for multiple instruments? Are there traps and filters to purify the gases? How do the gases enter the GC – electronic pressure or flow modulators?
- Where does the gas go within the GC? *Via* an autosampler or through to the inlet? Trace the gas lines.
- What types of autosamplers are there? What automated sample preparation and injection techniques can be performed?
- What types of inlets are there? What techniques can be performed to introduce the sample to the column?
- What types of columns are installed or available? Are there pre-columns, restrictors, retention gaps or splitters? How do they connect?

Gas Chromatography-Mass Spectrometry: How Do I Get the Best Results?
By Diane Turner, Mathias Schäfer, Steven Lancaster, Imran Janmohamed, Anthony Gachanja and Jason Creasey
© Diane C. Turner, Mathias Schäfer, Steven Lancaster, Imran Janmohamed, Anthony Gachanja and Jason Creasey 2020
Published by the Royal Society of Chemistry, www.rsc.org

- Are there any other GC detectors installed and if so what types?
- Can the GC be controlled through the keypad and/or software?
- How is the GC hyphenated to the MS? What controls the transfer line, the GC or the MS?
- What ionisations techniques are available on the MS? Only electron ionisation (EI) or chemical ionisation (CI) too? If CI, which gases are connected for this?
- What type of mass analyser is the MS? Single quadrupole, triple quadrupole (QQQ), ion trap (IT), time of flight (TOF), orbitrap?
- What parts of the vacuum system are there? Locate the mechanical pump, is it oil-filled? Is the high vacuum pump a diffusion pump or a turbo? What GC flow rates can it handle? What do all the pumps sound like?
- What cables are used to connect the different parts of the GC-MS instrument for communication (remote start cables)? To the PC? To the power supply? Don't forget the autosampler too and don't disconnect any cables!
- Where is the PC located and how does it communicate with the instrument?
- What software is used? Is it the same for both acquisition and data analysis?
- Are there multiple software licences so that data analysis does not have to be performed in the lab? If so, how do you get your data to it?

A lot of information can be gained from the laboratory, but instrument manuals, the instrument order or packing list and even internet searches can be helpful in gathering as much information as possible about the instrument, before starting to use it.

It is usual to have a purpose, for example an application, for the first time you will use it, but it is important not to rush the first method, even with previous GC-MS experience, otherwise mistakes can be made and even instrument parts broken. Chapter 1 discusses the importance of first fully understanding why a sample is to be analysed and to ensure that it is suitable for GC-MS analysis. Chapter 8 discusses method validation and the importance of finding out what requirements there are for the analysis, in terms of method performance, as this will help to determine how extensive the method needs to be.

Next, is to break the method down into manageable chunks, determine each of the steps and instrument modules for the entire method, including all inputs and outputs. Chapters 1 to 5 discuss sample collection, sample preparation, sample introduction, analyte separation then detection with GC detectors and mass spectrometers. Chapters 6 to 8 discuss the data analysis method and parameters, for qualitative and quantitative analysis, including a chapter on mass spectral interpretation. Alongside how each of these instrument modules work, are the practical instrument parameters, how to first choose the parameter value (rules of thumb) and then optimise the values. It is strongly recommended to create a plan for each step/module of how to develop the method. The hardware parameters/consumables to be selected (*e.g.* carrier gas type, column or an inlet liner) and the parameters to be entered into the software (*e.g.* temperatures, flows) and the optimisation for each step or module (*e.g.* range of parameter values to be investigated) should be included in the plan.

Once the method has been developed, it needs to be tested and the data analysed. Chromatograms can be integrated and method performance values can be calculated, as discussed in Chapter 6. Mass spectra can be interpreted, as discussed in Chapter 7. The samples can be quantified as discussed in Chapter 8. Once both the data acquisition and data analysis methods are completed the method must be validated, even if it is just to check repeatability and calibration curves, as discussed in Chapter 8.

Part of the method validation is to determine the frequency that instrument parts and consumables need to be cleaned and replaced. GC-MS maintenance is discussed in Chapter 9. When the new method is working well, it is recommended to prepare for problems by gathering details on the performance of the instrument, for example the vacuum level, baseline level, chromatogram, so that if there is a problem it can be quickly identified and fixed, as discussed in Chapter 10, troubleshooting. All maintenance and troubleshooting that is performed must be recorded in a log book for that instrument.

GC-MS is a sensitive, accurate, robust and useful technique, but care must be taken with the selection of samples to be analysed by it, the development of the method and the maintenance of the instrument for continuous use for the many analyses over the huge range of applications that it is capable of. We wish you luck with your GC-MS instrument and sample analysis!

Subject Index